30

AUTOMATION AND BEHAVIOUR

A Social Psychological Study

AUTOMATION AND BEHAVIOUR

A Social Psychological Study

J. K. Chadwick-Jones
Director, Occupational Psychology Research Unit,
University College, Cardiff

WILEY - INTERSCIENCE
a division of John Wiley & Sons Ltd
London . New York . Sydney . Toronto

Library of Congress catalog card number 74–99054
SBN 471 14300 6

Printed in Great Britain by C. F. R. Woodward (Bath) Limited Bath, Somerset

To
Araceli

Acknowledgements

The main field investigation for this study took place in 1956-57 and 1962-63 but further observation and follow-up interviews continued up to 1966. The Department of Scientific and Industrial Research and latterly the Social Science Research Council gave financial aid. The Human Sciences Committee and Automation Panel of the Social Science Research Council afforded valuable opportunities to discuss some of the findings at two inter-university seminars. Colin Sheppard, Ann Beecher and Claire Swire-Griffiths of the Department of Psychology, University College, Cardiff, assisted with observation of production crews and interviews with workers' wives.

At all stages of the research I benefited very much from the advice of Dr. Michael Fogarty of the Economic and Social Research Institute, Dublin, formerly Professor at University College, Cardiff, and I am especially indebted to Mr. G. L. Dennis, Group Director of Personnel and Social Policy, British Steel Corporation, for his active interest in the research programme.

Professor George Homans of Harvard University and Professor Tom Lupton of the Manchester Business School, made criticisms and recommendations which led to the development and extension of the initial draft. Dr. A. T. Welford of Cambridge University and Professor W. T. Singleton of Aston University discussed a number of points in detail and offered suggestions which were of great assistance. I am also grateful to Professor Robert Guest of Dartmouth College, New Hampshire, and Professor Albert Cherns of Loughborough University of Technology for commentaries which had some influence on the final draft.

During my stay in South Australia, I was fortunate in receiving encouragement in completing this work from Norman L. Munn, Professor of Psychology at Adelaide University.

Harold Christie and Graham Willoughby produced the diagrams and figures. The photograph in Chapter 6 of the Cleaning Line in the new plant at Port Kembla is reproduced by courtesy of the Broken Hill Pty. Co. of Australia, and the photograph of the Electrolytic Tinning Line in Chapter 11 by courtesy of the British Steel Corporation.

I wish to thank Mrs. K. Blackwell and Mrs. M. Woodman for typing the draft and final manuscript efficiently and with dedication.

JOHN CHADWICK-JONES

CONTENTS

Chapter 1

Introduction

Nature of the investigation

This is a research study. It is also a 'before-and-after' comparison of the attitudes and behaviour of workers involved in a massive industrial change which culminated with their transfer from a plant where the operations were carried out by hand to one using automatic controls. These were the personnel of the tinning section of the steel industry which had formerly consisted of numerous batch-production factories and which was now converted to a continuous-flow process, concentrating all tinplate manufacture in a small number of new plants. The men had worked in a production system which relied on manual and muscular skills of its operatives; in the new system the physical work-load has disappeared and the skill content and routines of jobs have radically altered.

The post-war development of new techniques in steel production in Great Britain has been rapid and has required action on the part of management and unions in order to avoid widespread unemployment. The sites of the new plants were located so that employees of the obsolescent factories could be transferred with little material inconvenience to them, and certainly without moving their homes. Since the extreme problems of protracted unemployment did not arise, the consequences of technological innovation which we are to examine can be considered as problems of the middle range: those of employees remaining within an industry through a period of major innovation and who have moved from jobs requiring their direct physical intervention to the monitoring and servicing of machine controls.

However, effects of this type of transfer and displacement from one type of job to another have tended up to now to be concealed from general observation by the sheer fact that they occur in the individual remaining in his employment. The changeover will have produced its major effects in forms which are *internal* to the individual insofar as it has modified the range and variety of rewards which he derives from his job. It will have had effects in the behaviour required of the individual in carrying out his technical role;

it may have introduced different social relationships or changed the expectations held of him by colleagues or supervisors. There are the new skills, conditions and status of jobs and there may be far-reaching changes in the social organization of the work group and its connection to other parts of the hierarchy of the firm. Thus, not only the content of jobs but the roles of operatives, supervisors and management may all be affected by the changeover to automated production. Another aspect which we shall deal with concerns the steel workers' leisure and the extent to which the in-plant changes may be reflected there.

The tinplate section of the steel industry has not previously been studied. Some six thousand workers are employed in producing the entire canning requirements for the British market and a considerable share of overall steel product exports. Innovation was more intense and more drastic than in steel-melting and ingot production. Output of tin-coated steel has been estimated as being some ten times greater than before the introduction of automatic controls.[1] It would be hard to find points of resemblance between the old industry and the new except in the nomenclature of a very few production jobs and in the superficial appearance of the product. The technology has become immensely refined by the coordination of continuous machine operations. Coiled strip in large quantities passes without interruption through a series of multiple rolls, gaining speed as it becomes thinner. The strip is re-coiled on leaving the rolls and transported by machine for chemical treatment in a series of further continuous operations until it is finally coated with tin by automatic methods. Formerly the rolling of steel bars and their surface coating was performed in separate stages on small batches lifted and manipulated by operatives and the machine aids were relatively undeveloped.

Basically the description and analysis which follows is an examination of the interview responses of two comparable groups of steel workers of identical work experience or background. But their attitudes and opinions are examined in the context of a study of the technical operations and of work behaviour which will be the subject of a detailed analysis in the following chapters. In this book the inter-relationships of attitude and work behaviour are explored; at the same time a body of findings is presented which shows in what respect those who have been transferred to the automated plant have gained or suffered in the process of industrial change.

After the transfers, what were the men's perceptions of the changes? If there was learning of new skills, of what did these consist and at what level of difficulty? Were there particular individual problems induced by the replacement of manual tasks by other skills? Did the men adjust easily? How far was their previous work experience helpful or otherwise? Did age affect the learning of automation skills? Interviewees talked in great detail about their technical and social environment, and from the observational data I also was able to use corroborative information on the regularities or routines of this environment. Their job satisfactions were the immediate focus of the interviews. Attitudes held and expectations formulated in the pre-automated situation gave a base for the study, against which the behaviour observed in the automated plant could be assessed. The strength and quality of opinions before and after the transfers could be compared and the men themselves would be asked to express comparisons on a number of areas of behaviour.

Research procedures

There have been two main sources of information about the human effects of the innovation: interviews and observational methods. Before the transfers took place interviews were held with managers, supervisors and a sample of employees; in addition a period of observation extended over eighteen months of task performance and interpersonal behaviour inside the factory. The transfers were staged over a period of more than a year to a new plant which was already commissioned and operating with a minimal complement of workers. Six years after the first of the transfers, an equivalent number were interviewed, using the same method and the same questions but this time also asking for comparisons on specific aspects of jobs. Observations were again made of work performance and group behaviour, and on this occasion certain production teams were under continual observation for periods totalling several hundred hours and communication networks were analysed.

Before the interviewing programme could begin, talks were held with top management and union officials to discuss the general purposes of university research. It was emphasized that information could best be obtained on the human effects of changing technologies if those who were involved expressed their own experience of it. The companies and unions concerned offered to assist in the investigation, and I discussed the project with plant managers and supervisors and attended union meetings to give further explanation. The employees, who were asked to help in the data collection by answering the interviewer's questions, gave their views at length and interviews of two or three hours were not unusual. There were no refusals in the interviewing programme. No interview took less than one hour and the longest lasted over six hours (taken in two-hour daily instalments).

The method used was the depth or free-answer interview, followed by a schedule of questions on specific aspects of the work environment. Interviewing began with questions on biographical background leading on to information on age, length of service, work history and the title and location of the interviewee's present job. He was then asked for a self-assessment of his degree of liking or disliking the job, and the interview subsequently entered the phase where open-ended questions were asked: in this part of the interview, after an introductory general question, the topics mentioned were only those spontaneously raised by the interviewee. This was the principal value of the free-answer interview approach—starting with the question *What are the things you like or dislike about this job?*—it encouraged the interviewee to talk about the positive or negative features which he himself considered to be most important.

Questions on a number of specific topics followed later when the free-answer information had been collected, and the interviewee indicated that he was ready to continue with more questions. In the interviews after the changeover the employees were asked to make the comparison between different aspects of their present and previous work situations in terms of preferences and also to give their views on what for them were the main or most important effects of the technological advance.

Arrangements for interviewing the employee sample inside the factory, the time order of interviews and the substitution of employees in the work teams while they were interviewed, were made by the management of the departments concerned, after they were given lists of names of the sample. It was inevitable in this project, extending over a long period, that many conversations and discussions with employees took place in chance meetings in the urban community where most of the interviewees lived, in meetings of the union branch and at the homes of workers and managers. It was even difficult when the project began to limit the interviews to those selected from the pay-roll lists as the one-in-five random sample of different grades. Others expressed an interest in being interviewed, approaching me and offering information: 'I thought you would like to know . . . '. Further interviews were held with those who were referred to me by their work mates because they 'ought to be interviewed' and would make a good contribution. Although this larger, and as it turned out, cumulative population of employees was interviewed, the numerical analysis on which the description of the change is based does not include the total number. However, these additional interviews gave further illustration to that provided by the sample. They added weight or confirmation to the results and if they had been included in the analysis they would have supported the trends of opinion which the sample data revealed.

Sampling: Conditions of industrial investigation

The random sample, taken before the changeover, was stratified for the different grades of employees and comprised forty-seven operatives and ancillary workers employed in the rolling and surface treatment of steel. In the great majority the workers were semi-skilled by industrial definitions—that is, they were not required to undergo formal training for their jobs but some learning of special skills and acquisition of knowledge by job experience was necessary. They were interviewed as part of a wider sampling[2] from the various sections of pre-automated steel-making: furnaces, melting, bar mills, rolling and tinning departments during 1956 and 1957.

After the introduction of the automated system fifty-one employees were interviewed in the new factory at the end of 1962. This was now after an interval of five or six years since their transfer and re-engagement. This number was interviewed because they comprised *all* those transferred from the old factory to a new plant, which specialized in rolling and surface treatment. Other employees of the old plant were also re-engaged within the industry but they were dispersed in small numbers around different plants—for example, in labouring and ancillary grades in ingot production—and outside the tinplate division of the company. A more extensive research programme to include them was not practicable.

To summarize: before the changeover, interviews were held with a sample of those engaged in rolling and tinning operations; these operations were superseded by a continuous machine system and those who were transferred to it from the old factory were now interviewed. Like the interviewees in the old factory, they were mostly married; all lived in the same township and were of similar age distribution to the previous sample, whose mean age was a little over forty-six years.

The members of the 1956-57 sample were representative of all the plant population in rolling and tinning operations before the changeover. The population of the new plant in its entirety was not represented by sampling—which would have required a larger and random selection from the various departments. For this reason the second run of interviews under-represented the increased proportion of maintenance personnel employed in the new technology.

However, a study of employee record cards showed that in their industrial experience and community background the interviewees were not differentiable from the general plant population who had been recruited in most cases from the obsolescent factories of the steel industry and were drawn from local urban communities within a region of twenty miles. They tended to be older, although this was not a very marked difference: the mean age of the interviewees was 47·7 (S.D. 10·3); the mean age of the entire factory labour force was 43·1 (S.D. 11·95).

The men were interviewed in their homes as a further check in the autumn of 1963 by a research assistant. Again the material confirmed the results of the interviews in the factory and was therefore not treated with a separate analysis. It was of interest principally for the replication of the trends of opinion revealed by the in-plant interviews. Wives of the men were also interviewed at their homes. Thirty of them who were already married during the period of their husbands' employment in the old factory were asked to give their opinions on the results of the transfer.

One of the senior managers, expressing his views to me before the second investigation, said that he would welcome 'anything you can do which may lead to improving our industrial society', and this statement reflected the extent to which the management were helpful: facilities for the research were willingly provided. But there were restrictions on the research programme because of conditions laid down by the directorate of both companies concerned. For example, after the first interview run, a questionnaire distribution was proposed to cover the employee population. It could not be allowed by management because of the fear that it might increase rumours about the limited future of this production system. In the new factory permission was sought to interview a representative sample of all employee grades, but this was not agreed to because of a recent wage dispute with the unions in the company. It was a choice between carrying out the study under these conditions or not at all.

Contents of the following chapters

In chapters 2 and 3 analysis is made of tasks, skills and interpersonal behaviour in the production teams of the old plant. The autonomy of the team is revealed in controls exercised by its members over work pace and alternation of tasks, and the immediate rewards available to the individual from collusive activities are described. Insofar as these also modified the technical demands they can be termed *social regulators.* The local union is shown to have power over the structural composition of teams by filling vacancies and implementing requests from individuals for promotion or transfer.

Connections between community and factory and their influences on factory behaviour in the period immediately before the innovation are the subject of Chapter 4. The extent of kinship and neighbourhood ties and their effects on formal discipline are considered. Residence in the community led naturally to employment in the steel plant. Innovation effects can be more adequately understood against these factors which are referred to as *external* to the industrial situation only for convenience of analysis.

A summary of the attitudes of interviewees before innovation is presented in Chapter 5. This provides the base for subsequent comparison when the analysis is repeated for attitudes in the new jobs. The concept of satisfaction is considered in relation to the constraints on individuals to acquiesce in a situation of limited choice.

The new technology is described in Chapter 6, and here automation taxonomies provide a background for an examination of the changed tasks of production workers. Analysis is presented of the jobs of operatives in the electrolytic tinning section of the plant where a majority of interviewees were employed. Clerical tasks, machine setting and fault detection are detailed, and the chapter ends with a discussion of the difficulty level of jobs.

Features of job and environment which attracted positive attitudes are discussed in Chapter 7, and in Chapter 8 the conditions or topics associated with dissatisfaction in the new plant are dealt with. Changes in some characteristics of the jobs produce what is termed as a *reversal of satisfiers* where they become, for example, the focus of favourable instead of unfavourable attitides. The main topics of Chapter 9 are the responses to direct comparison questions between conditions in the old and new plants. The process of an individual's adjustment is analysed in this chapter.

Other post-changeover effects and particularly the separation induced between leisure activities and occupational affiliations are discussed in Chapter 10. Recruitment to the new plant was from a number of different urban areas and shift working made it unlikely that previous and traditional leisure activities would continue. This had the effect of reducing the social cross-ties and more extensive relationships formerly present in the workshop. In the employees' domestic life a new pattern is described, and interviews with wives show a greater emphasis on the home and the immediate family in the use of leisure time.

Inside the new plant there was a transformation in the structure of work teams on the production line. Chapter 11 gives a description of both attitudes and behaviour which makes it possible to compare them with previous modes of social behaviour. Union activities had also taken different forms and this was of significance for the primary group.

Touching on these changes were relationships with supervisors and managers which are described in Chapters 12 and 13. Supervisors' behaviour towards their subordinates is described together with the interviewees' reactions. A similar examination is carried out on managers' behaviour, and answers are given to the question of whether there are differences between employees' expectations and the actual behaviour of supervisors and managers.

Findings and concepts

The notion that the individual attempts to reduce dissonance between his expectations and what the new situation has to offer, is used to explain acceptance of the jobs in the new plant and adjustment is seen to consist of an anticipatory stage followed by task learning and a much longer process of learning new social roles. Stress is defined especially in relation to the uncertainties of the situation of innovation but it also appeared in the context of social relations in the new plant.

Changes in the primary social group presented difficulties for the individual, and there may be similarities in this respect to conditions which have been observed in the interdependent elements of structure (whether in engineering, physiology or neurology) where over-large variation in one or more of the elements produces eventual structural disintegration. For example, in physiological systems the precise limits can be specified of *essential* variables which are most closely related to survival of an organism.[3] In other words, the analogy suggests that in social behaviour certain variables may need to be held within given limits if social patterns are to undergo change without breaking down to the point of stress for the individual. This is not at all the same as stating that there is a particular virtue in stability as such.

Let us look at this point more closely. In physiological systems there are degrees of change in variables which are at the same time consistent with general continuity of structure, and any condition of activity incorporates or expresses this type of change. Some of the variation involves regulating or compensatory mechanisms which maintain normal functioning. But it may be that a change in a component or element is of such an amount that it cannot be affected by any counteracting mechanism. This result, producing a complete breakdown in structure, might occur during very large or rapid environmental changes. Some variables can modify to very wide limits without affecting normal functioning. On the other hand, others which are basic and essential to the structure may need to be maintained within relatively narrow limits, and so long as they are it may still be possible to allow very considerable modification or even substitution in other components.

If we were to attempt to apply this analogy further to social behaviour we would find that here both the essential variables and the degree of permissible variation are largely a matter for conjecture. They might be any of the following: some existing attitudes, standards of behaviour and the range and frequency of shared activities in a social group. These are variables which are important for form and structure in interpersonal networks.

It is necessary to emphasize again that it is not the maintenance of generally stable conditions which is important but the need to bring to light significant elements of behaviour in social groups and to use this knowledge in order to innovate towards optimal states. In order to decide between lines of action or experiment, assessment of a situation will first have to be made so that it can be discovered where change is likely to be most rapid or inclusive and which behavioural variables will be modified. At this stage decisions might be taken to maintain certain known social routines or to substitute them with others. It will be necessary to recognize that some of the existing social forms might provide the employees with sufficiently stable points of reference to allow them to cope more easily with other major changes, for example in learning unfamiliar skills. Once these alternatives are recognized and defined, estimates of preferred social group structures might be included in forecasting, not as a means to adapting human behaviour to innovation but as an approach to innovating both in behaviour and technology. These possibilities will be referred to again in the final chapter, where they will be discussed speculatively. There are several issues involved in such a discussion: the exploration of theoretical relationships between the variables; their importance for the individual's well being; and their relevance for organizational planning.

Comparable Studies

Two studies had already been carried out on technological change in the steel industry, with contrasting orientation and methods. Both were studies of 'after' situations. The first was a case study of a production team in a pipe rolling mill by Charles Walker of Yale University, where the operatives were interviewed at intervals of two or three years after the introduction of new machines.[4] The other, by W. H. Scott, J. A. Banks, A. H. Halsey and T. Lupton of Liverpool University, was a general survey of the effects of change on industrial relations in steel blast furnace and melting shops.[5] Neither of these situations involved highly advanced levels of automation. Walker emphasized in his study that the technology was at a stage 'towards automation'. However, this did not detract from its importance as a fascinating account of a team of operatives coping with the transformation in their jobs. In the Liverpool study, which was also of an earlier technological stage, some of the results are of direct relevance, particularly the effects of new shift arrangements on leisure activities.

REFERENCES

1. W. E. Hoare, E. S. Hedges, B. T. K. Barry, *The Technology of Tinplate,* Edward Arnold, London, 1965.
2. Some of the material from the larger sample is treated in J. K. Chadwick-Jones, 'Managerial Leadership in a Steel works', *Journal of Management Studies,* 2, No. 3 (1965).
3. W. Ross Ashby, *Design for a Brain,* 2nd Edn, Chapman and Hall, London, 1960, p. 42.

4. Charles R. Walker, *Toward the Automatic Factory,* Yale University Press, New Haven, 1957.
5. W. H. Scott, J. A. Banks, A. H. Halsey, T. Lupton, *Technical Change and Industrial Relations,* Liverpool University Press, 1956.

Chapter 2

The Pre-automated Technology

The production of tin-coated steel (tinplate) in the old plant was carried out in a sequence where batches of steel bars or sheets were manually transported between each of the various rolling and surface-treatment operations before the final coating with molten tin. A principal observable characteristic of the production system was the activity and movement of groups of operatives lifting and positioning the raw materials near furnaces and powered-machine rolls. The physical effort required of the operative in environmental conditions of high temperature and radiant heat was the most obvious characteristic of the work, but there was more to it than the physical activity alone as I shall try to show in this and the following chapters. There were operations which required the exercise of judgement and discretion by the operative; above all, the work environment was socially complex, in its alternations and exchanges of jobs between individuals and in the procedures formulated among the workers for their coordination.

Technical, individual and social components

The classificatory framework used to analyse employee behaviour in a factory should take into account the technical, individual and social components of work organization because these all influence, in some degree, the forms of behaviour which appear in a given situation.

It may be assumed that the work performance of the industrial employee is a resultant of these interdependent components: the technical demands, his own capacities and the work roles, duties, obligations and rewards accepted and established in his social group. The nature of the pressures or demands of the technology on the worker must be considered in relation to the repertoire of his (individual) responses which they elicit; but these may be simultaneously modified and the technical demands filtered or controlled by collusive (social) responses. The point is worth emphasizing that it would not be adequate to carry out a study of opinions or attitudes at a level too far removed from the behavioural context of work. In the case of our steel workers, their attitudes had evolved from a particular technical and social environment and existed in relation to various events occurring in this environment, and these have to be described.

The technical background

The operations at each production stage involved teams of six to eight operatives working in close proximity to each other, intervening continuously in the sequence or cycle of operations and frequently exchanging their tasks and positions. These features were most clearly illustrated in the rolling teams or crews: over half of the sample were production or ancillary workers in this section of the plant.

The rolling mills provided a spectacular first impression. Groups of three or four men moved and turned near a line of roll-stands four feet high through which they passed, re-passed and swung from one to another, sheets of red hot steel weighing up to forty pounds. As many as eight crews could be seen working in a line, adjacent to each other, pushing the sheets along the smooth metal floor near the roll-stands, each crew working within an area less than twenty feet square and all following a similar pattern of movement.

The work team

The full crew for a set of rolls was made up of six operatives in a wage-grade hierarchy consisting of rollerman, doubler, furnaceman, first and second helpers, and back-roller.

Figure 1 shows the working area; the team worked between two roll-stands and two furnaces (one serving each set of rolls). The roughing rolls were used for the preliminary rolling of steel bars, the second stand held the finishing rolls, a finer gauge pair, for use on the sheets in the later sequences of the work cycle. The furnace for the roughing pair was used to heat steel bars from their cold state on entering the mill and the finishing furnace for reheating and maintaining the sheets at a constant temperature between the rolling operations. In the course of rolling steel bars from $\frac{7}{8}$ in. in thickness to layered sheets of less than $\frac{1}{16}$ in., operatives used hand tongs of four different sizes; with these they positioned the bars and sheets into the rolls or arranged them in the furnaces. The tongs ranged from two-and-a-half pounds and five pounds in weight for the rolling operations to seventeen pounds in furnace work.

The men were exposed to temperatures of around 100° F (38°C) near the steel sheets and furnaces[1] and the physiological pressures were onerous to the extent that those in rollerman, doubler and first helper grades invariably referred to the necessity of one or two hours' sleep at the end of the shift.

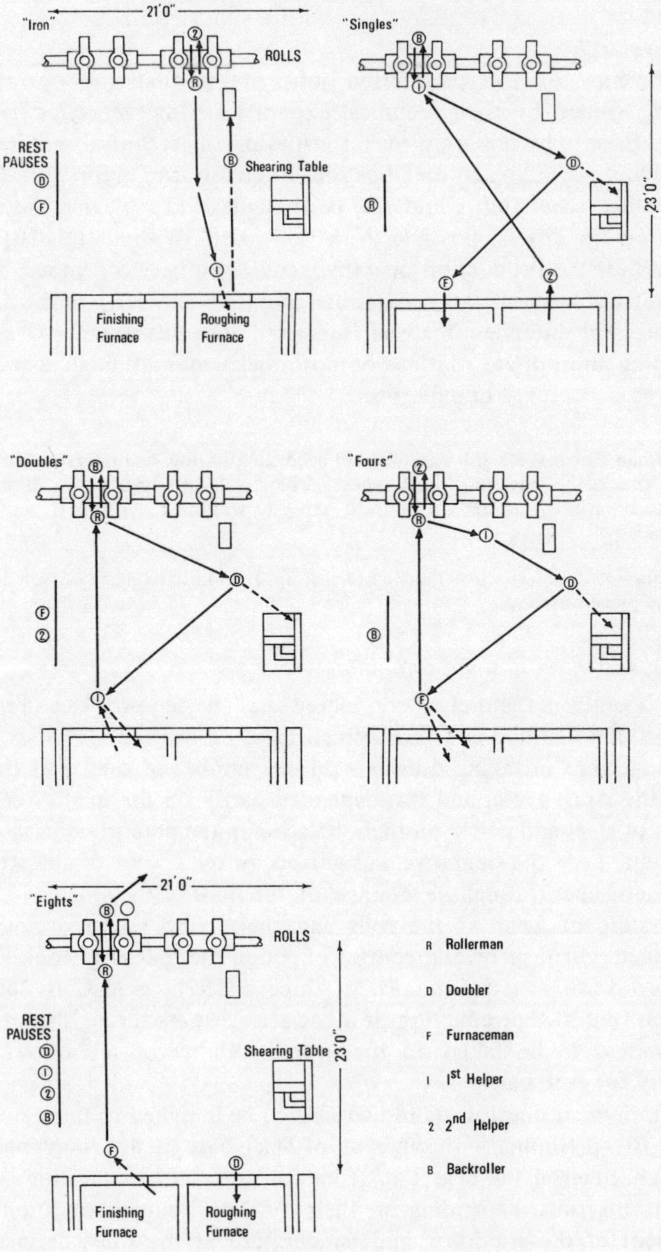

Figure 1. The five parts of the work cycle.

Operative skills

Proficiency in the production jobs meant that the operatives took decisions guided by their technical experience. However, *effective* performance in these jobs not only meant achieving the technical requirements for the product according to specifications of quality and evenness in the surface of the rolled sheets but could also be thought of as achieving this result at as low an energy cost as possible to the individual. Without this latter type of effectiveness the production operatives could not have continued, as they did in all but the heaviest tasks of doubler and first helper, up to the age of sixty and over. The interview material suggested that this was partly a matter of developing appropriate routines of motor behaviour although, as we shall see, there was also a range of expertise:

'A rollerman can make a job easy or hard according to how he understands the rolls. We call it a knack, the way you flip the sheets. There is definitely an art in all the jobs. You can make it easier or harder for yourself and you wouldn't last long if you didn't have the knack'.

'The experience counts a lot, that lightens it up. I could lift a piece of iron and it would feel like a piece of wood'.

Decision and discretion in production workers

The operatives themselves considered that the senior jobs and particularly the tasks of rollerman and furnacemen involved most art. The six operatives used judgement in taking decisions on the number of steel bars to be rolled during the work cycle, and this depended partly on the quality of incoming batches of steel and partly on their variation in temperature during the rolling operations; here the operative was guided by the colour of the steel and by kinaesthetic cues through its 'bounce' on the floor of the mill.

The state of wear of the rolls and their temperature during the shift determined which of several courses of action the rollerman took in applying the screw-down mechanism; two or three different gradations of pressure were involved in the reduction of a bar at each pass through the rolls, and a decision had to be taken on the overall number of reductions or passes necessary for each bar.

Operations in one roll-stand had also to be matched to those in the other. During the preliminary throughput of steel bars in the roughing rolls, the rollerman changed the degree of reduction to suit the conditions of wear in the finishing rolls, depending on their 'full' or 'hollow' condition; from his assessment of the condition and temperature of the rolls, he informed the furnaceman of any temperature changes required in the bars or on the speed of their withdrawal from the furnace.

The furnaceman's duties included vigilance of some forty bars or sheets inside the furnaces; these had continually to be rearranged and moved in order that they might be of uniform temperature when taken out for the rolling sequences, as well as to ensure that the increase in temperature of the bars during the work cycle was gradual. Obviously, the furnaceman's tasks were directly coordinated with those of the rollerman during the whole cycle of operations.

The direct interdependence of work roles which we could observe between rollerman and furnaceman was characteristic for all the positions of the crew as was interchangeability of tasks. This will become clear if I now describe the rotation of tasks which was repeated *during each hour* of the work shift.[2] The sequence of operations for each of the five parts or stages of the cycle is illustrated in Figure 1, which shows the pattern of standardized formal interaction, i.e. the work contacts and basic routines required in each team.

The five parts

The hour-long cycle or sequence of reducing the steel bars into eight-layer sheets was divided into five parts of ten to fifteen minutes each. The arrows in Figure 1 indicate the direction of movement of the bars or sheets. The arrangement of rest pauses for each member of the crew is also shown. In the first three parts (iron, singles, doubles) these pauses lasted for about ten minutes, i.e. corresponding to the duration of the parts; in the last two (fours, eights) they were usually five minutes longer.

Rest pauses

This arrangement of rest pauses compares well with the recommendations of a study by Scholz, who assessed the appropriate rest periods for foundry employees working in high temperatures at fifteen minutes to every forty-five minutes of work.[3] An examination of Figure 1 shows that most members of the crew took a rest of ten to fifteen minutes during each hour, while for the furnaceman and doubler, the period was for twenty minutes duration or more. Informal exchanges of jobs introduced further breaks, particularly for the rollerman, and for all members of the crew occasional early leaving and extra rest pauses were achieved by this means.

It is not possible to give an adequate idea of the team operations without describing in detail what was really a very intricate pattern of connected roles even before the informal exchanges of tasks superimposed additional patterns, which will also be described. The sequence must be given detailed treatment because it formed the basis for the elaboration of social norms and for the customary procedures of behaviour in the factory.

The Sequence of Team Operations

Part 1–Preliminary rolling of bars

The back-roller (B) took the bars out of the roughing furnace, placing them on a stand beside the rollerman who was in front of the roughing rolls. The rollerman (R) passed the bars through the rolls three or four times each, increasing the pressure after each pass by means of the screwing mechanism. The second helper (2) stood on the other side of the rolls passing the bars back over the top roll. The bars, now having been reduced in thickness to the consistency of sheet, were thrown by the rollerman back towards the roughing furnace. The first helper (1) lifted them into the furnace, three feet from the floor of the mill.

Part 2–Single sheets

The second helper threw the sheets out from the roughing furnace across the mill to the first helper at the finishing rolls. The first helper passed them through the rolls twice, with the back-roller on the other side to return them over the rolls. The first helper then threw them down towards the doubler who folded the sheets before passing them to the furnaceman.

In the operation of folding or doubling the sheet the doubler raised one end of the sheet from the floor, brought it down to meet the end still resting on the floor and used the pressure of his foot to bend the sheet double and to hold it in position while he gripped both ends with his tongs. The doubled sheet was then lifted to the table two-and-a-half feet high on the right of the mill where it was pressed and cut by a mechanical shear.

During this part the doubler and second helper also recharged the roughing furnace with bars in preparation for the next work-cycle.

Part 3–Two-layer sheets: 'the doubles'

The doubled sheets were thrown down from the furnace by the first helper, towards the rollerman. The rollerman passed them through the rolls with the back-roller returning them. After two passes the rollerman tapped the corner of the top sheet forming an 'ear' by which he held the piece (the term used for the multi-layered sheets), throwing it down towards the doubler and at the same time separating the two sheets, to prevent them welding together. The doubler folded the piece making four sheets, aligned them and lifted them on to the table for trimming and cutting at both ends, before passing them to the first helper for replacing in the finishing furnace.

Part 4–Four-layer sheets: 'the fours'

The four-layer sheets were thrown down from the furnace by the furnaceman towards the rollerman; the second helper passed them back over the rolls. After two passes the rollerman tapped the corner so that an 'ear' was formed on the two top sheets. Holding this he threw the piece down, separating as he did so the two top from the two bottom sheets. The first helper then opened the top and bottom sheets and gave them to the doubler for folding, pressing and cutting. The doubler threw the piece towards the furnaceman who lifted it to the finishing furnace.

Part 5–Eight-layer sheets: 'the eights'

At the beginning of this part the doubler turned the bars that were placed in the roughing furnace at the end of the singles, placing them on their edges on both sides of the furnace so that they were heated evenly, at the same time making room for the singles to be replaced during the first part of the following cycle.

The eight-layer sheets were now taken out by the furnaceman. The rollerman rolled some of them with the back-roller and the remainder with the second helper.

During this part the rollerman could check on the length of the sheets with a measuring rod. The back-roller and second helper piled the eights outside the mill. This was the final operation of the cycle and one of the most spectacular, the operative swinging the sheets out of the rolls, bouncing them on the floor and lifting them onto the pile (on the left of Figure 2), all in one continuous movement.

Direct intervention of the operatives

The completion of the process depended, therefore, on a pattern of physical intervention and manipulation of materials by the operatives. The technical interdependence of the crew was not lessened by the fact that judging and discretionary action was taken mainly by the rollerman and furnaceman, and the skills of the remainder of the team were very largely those of muscular dexterity and coordination. Even the back-roller or the second helper, although their duties were lightest in the crew, could through faulty technique make errors affecting the quality of the product. For example, if the back-roller shook or jerked the layered sheets as he returned them over the rolls, air was allowed between the layers causing them to stick together and damaging the surface. A similar mistake could produce this effect during the piling of the sheets outside the mill at the end of the last part.

The operative was a direct link in the rotation of job operations among the team members, and even though the physical tasks of, for example, the back-roller might be of a relatively simple nature his intervention was

Figure 2. The final operation of the work cycle. In one movement the back-roller swings the eight-layer piece from the rolls, bouncing it from the floor onto the pile in the left of the photograph.

Figure 3. The Mill Crew. *Left to right:* First helper, back-roller, rollerman, doubler, furnaceman, second helper.

essential in the sequence, as was pointed out many times by the senior operatives and expressed in the dictum that 'the back-roller is the rollerman's best friend'. So, each operative had immediate and easily executed forms of control over both quality and pace, and the sharing of complementary roles could in itself be assumed to have had an influence on this work group. A second helper discussed the contribution of the back-roller as follows:

> 'Every man is important in the mill. The back-roller is as important as the rollerman, he can spoil the work as much as anyone. A sensible fair man would tell you that every man is important and when you get every man doing his best it's easier for everybody. And if you get one man doing nothing it makes it hard all round. You take a back-roller in the mill. He's nobody in the mill, but as his work goes out, if he shakes them or is a bit rough or didn't put grease they'd "catch" and he'd spoil the work. The point is that everyone's job in the mill is important'.

Informal exchange of roles

We have described the sequence of operations and, in doing so, the rather elaborate interconnection of a series of shared and mutually dependent roles. It was also mentioned that there were further possibilities of spontaneous exchange of tasks. This meant that the sequence could be varied by substitution of one operative by another, if fatigue made it difficult for an individual to keep up to the pace of the crew or to perform satisfactorily his role in the work cycle. In other words, any operative in the crew could be a relief for other members, allowing extra rest pauses and even allowing a team member to leave at the start of the final cycle, cutting his shift by as much as an hour, his work being shared out between the remaining five.

Interchanges routinized in crews

Besides spontaneous occasional job exchanges between individual operatives there were also several types of task sharing which had become general and routine to most of the crews. Arrangements of this type were usually between the rollerman and members of his crew. The exchanges were as follows, taking them in order, as they occurred in the different parts of the work cycle: in the first part the doubler rolled seven or eight bars, the rollerman helping in return with positioning bars in the furnace; during the fifth part the first helper rolled the last four sheets for the rollerman; the doubler also rolled some of the eights and the rollerman would often double a few sheets in return. Where such exchanges involved crew members of a very different status in the operative hierarchy, for example the rollerman and second helper, then the senior man would maintain the status difference by rewarding the junior grade 'with a couple of beers'.

All these possibilities for job exchange had in common the fact that they depended on the nature of the relationship existing between members of the crew; they were voluntary and if personal antagonisms or friction arose between members they could be terminated: 'If we have a tiff, it's off!'. The value of friendship formation in the crews thus had as its material pay-off the reduced probability of this type of friction.

Figure 4 illustrates the pattern of contacts during each hour as they were required in the job activities of crew members, together with the elaboration produced by the almost routinized interchanges which have just been specified.

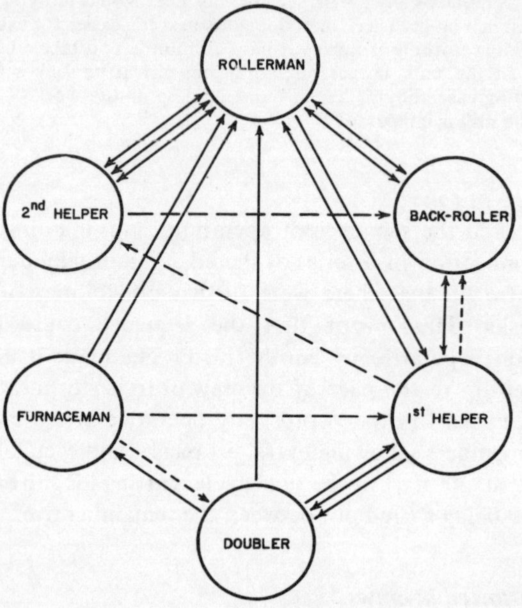

Figure 4. Work contacts per hour

Each unbroken line represents ten to fifteen minutes of the main task contact. Dotted lines represent the further contacts which resulted from additional task substitution. The more temporary exchanges in response to individual needs are not, of course, shown here but will be discussed again later. Even without the latter the pattern is one of intensive team interaction.

Autonomy of the operating crew

Yet another source of control over job demands derived from the autonomy allowed to the crew by the management of the company. Within an output limit fixed by union-management agreement, the crews were paid by group piece rate. They were, in fact, left to decide on the number of bars to be rolled during each hour: the rollerman usually taking this decision for or in consultation with his crew.

The throughput of bars and therefore the work pace for the crew was partly determined by the rollerman's assessment of such technical factors as the temperature and condition of the roll. The decision needed to be taken at the start of each hour of the shift, determining the number of bars to be passed through the preliminary rolling operations of the first part.

Again, his discretion extended to maintaining an hourly level of output which was also geared to the capacity of crew members to work at a given pace: in other words, taking into account the physical strain on individuals and, if necessary, cutting down on the throughput during the cycle of operations. The rollerman's leadership extended to 'judging' the capacities of the crew and to keeping a watch for signs of strain, 'in hot weather I'm supposed to be judging them, . . .'. The rollermen interviewed explained how they could take action to alter the pace: 'if one of the crew is "down" or not feeling too good in health then we work accordingly, therefore you have to slow down. If I see the doubler not so well, I'll cut the charge and make it less'.

This control by the operatives over rate of output, and the variation in speed which was possible during or between each hourly cycle, was the result both of the conditions of work in this type of batch production and of the loose degree of organizational integration in the company. Direction from outside the crew was not required because the operations were not technically refined to the point where they needed expertise other than that possessed within the crew, and supervisory checks were observed to occur infrequently.

The rollerman was accountable for his technical decisions relating to the state of the rolls, and for any faults in the quality of the product to the supervisor who checked on the output of the crews at the end of each shift. For the adequacy of his decisions, in a different sense, he was accountable to his own crew continually during the shift; he was required to assess the physical condition of individuals and to set the rate of output to suit them.

Social controls over the work process

Apart from the subtleties of the decisions required from the rollerman and furnaceman and the muscular dexterity of the doubler, the team tasks had a basis of repetitive actions. From this basis a number of variations had developed to produce a more complex pattern; but the source of complexity was social rather than technical.

Besides cutting down on throughput, controls in the contrary direction could be implemented in order to increase the speed of working. In the crews with younger members it was frequent practice for the pace to be increased to the point where the men could shorten the length of the shift for the entire crew by half an hour. These possibilities allowed great flexibility in relation to the work load which was allocated along a time dimension, by increasing or decreasing the pace throughout the shift and by changing even the duration of the shift itself.

Another result of the behaviour we have described in this chapter was to diffuse the load among members of the team, depending on the needs of individuals. Changing the content of the work tasks or the order of their performance, by these collusive actions, were forms of *social compensation* on the part of the various members but leading, obviously, to better coordinated functioning in the crew as a whole. The support roles exemplify an extreme development of norms for reciprocal activities, and these behavioural modes were more than affiliative in the satisfaction they yielded to the individual. Making this explicit: more than the general incentive of association with others the end-results were the actual rewards of variety in work, avoidance of fatigue or easier task performance and their incidence was both frequent and immediate.

A finding to be reiterated is that the interchanges were entirely at the discretion of the operatives in each of the crews. Moreover, there was no restriction on substitution of jobs to any part of the hierarchy. Although there was differentiation by wage level and task duties, from rollerman down, exchanges could occur between individuals of any grade. This also achieved a training function in giving practice or competence to an individual in jobs other that his own and it created an area of tasks shared by all grades.

Summary

Analysis of the technical, individual and social meaning of work behaviour before the changeover started with a description of the tasks of each operative in the hand mill crews. The nature of the skills; the decisions and discretion exercised by the senior operatives were considered. Some of the decisions were technical but others were social, such as the rollerman's 'judging' of effort and fatigue in members of his crew, with consequent variation of throughput of the mill. It emerged that the tasks were carried out in coordination with others to a high degree, and that another characteristic of the job cycles was that they were interchangeable between operatives. Figure 1 showed how the five parts or stages of the hour-long cycle required a complex of exchanges of position and task, with alternating rest pauses. The sequence of operations, the equipment used and the actions taken by the operatives were presented in detail. The importance of the operative's direct intervention was defined in relation to the possible results of poor performance; in this respect it was seen that 'everyone's job in the mill is important'.

In addition to the complex pattern of hour-by-hour task contacts and exchanges required in the production team, it was demonstrated that the sequence was subject to further variation: by additional relief duties or rest pauses. Some of these had become routine task-sharing general to the crews and added to the main pattern for the five parts they resulted in the hourly configuration of shared activity illustrated in Figure 4. These and other, more spontaneous, job exchanges were discussed and related to the autonomy possessed by the crew and its collusive actions.

REFERENCES

1. This aspect of work load is examined in R. K. Macpherson 'Physiological Responses to Heat Environments', Medical Research Council, *Special Report Series, No. 268.* London, H.M.S.O., 1960.
2. The job analysis was based on observation of task content and on information from the answers to interview questions aimed at assessing the nature and frequency of mill operations.
3. H. Scholz, 'Changing Physical Demands on Foundry Workers in the Production of Medium Weight Castings', *Ergonomics,* 1 (1957). Scholz describes conditions similar to those of the hand mills, particularly for heat stress.

Chapter 3

Behaviour Expectations: The Social Group Before the Changeover

The work groups before the changeover possessed clearly definable characteristics which will be detailed in this chapter largely from the interviews and observation inside the plant and, in addition, from a content analysis carried out on the records of the trade union branch for a twenty-year period. This helped to place the observed job behaviour in its time context. While it appeared to have well-established forms, the branch records showed that some of these had developed to their current state during little more than a decade before the changeover. Although an employee would occasionally refer to some aspect of his industry and community as existing 'from time immemorial', different and more autocratic patterns in plant behaviour were within the experience of the oldest employees (see Chapter 13). The norms which are described in this chapter had developed during the ultimate decade or two in the old factory to a point where they were particularly advantageous to the individual in this work system.

To understand the forms and purposes of behaviour in the production teams we will consider briefly the process of *informal* training of newcomers to the labour force. New entrants into the factory were taught in the work crews both the standards of job performance and the appropriate interpersonal action which was expected. This was the equivalent on the primary group level of formal training procedures in an industrial firm and the term 'social induction' could be used for it.[1] Moreover, the induction or training aspect of group behaviour became especially active during the last years of the obsolescent plant when workers were recruited in Italy on contract to the steel industry. For these workers, social training was only possible in face of obstacles which were an exaggeration of everything that might have occurred in normal cases of recruitment into the industry. However, such an abnormal situation did throw light on operant processes which usually would be much less available to observation; it will provide illustrative examples later in the chapter of the characteristic features of the hand mill social organization. In return for learning the appropriate behaviour, a number of benefits were available to the individual and we shall now examine their nature.

Interpersonal behaviour as a regulator

The description of production operations in the last chapter dealt largely with the technical requirements and the repertoire of sensorimotor or physical skills, but to describe the operations accurately it was necessary to refer to the ways in which individual responses were modified or guided by social cues. There were plenty of indicators that behaviour had developed in such a way as to provide *regulators* which could operate through the collusive roles of the crew. Figure 5 gives an illustration of the development and operation of the social regulators, acting on the technical requirements in the closed-loop fashion of a highly adaptive system. Routines of work were modified or corrected by the actions of the crew in response to the conditions experienced by them: roles were re-allocated and distributed at will to introduce a great variety in the task sequences which was essentially sensitive to individual needs or capacities.

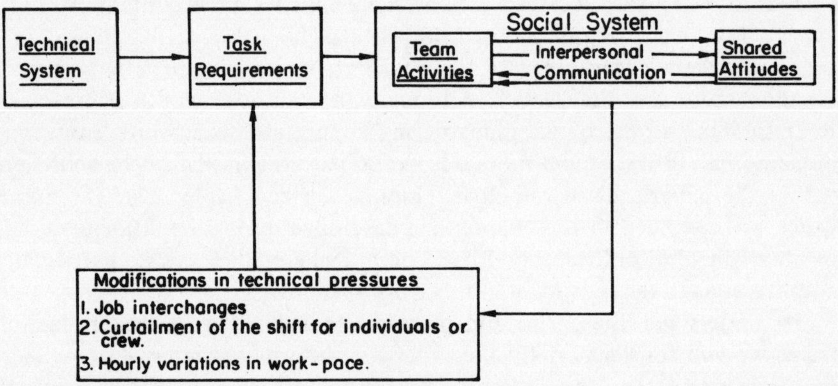

Figure 5

As we saw in the last chapter, each operative's job might involve, as well as his own tasks, some overlap with those of others, leading to an extension of his task area into that of other members of the team. The job operations of each man thus formed a series of mutually overlapping areas. There was no sharp demarcation of each man's role, but a large common area of tasks which might be shared or where the performance of one man might be substituted for another. The extension of a given operative's duties depended directly on interpersonal relationships in the team and therefore the pattern of the day's activities was related to arrangements between members.

It was made sufficiently clear in the interviews that the rewards to an individual in return for the extension of his work load, or for taking on additional tasks, were the reciprocal services and support of other members of the team. The primary group thus supplied its own reinforcement to its members for sustained levels of work performance.

The results for the individual were as follows: depending on his needs or preferences, exchanges of jobs or additional rest periods could be arranged, or the pace of work of the entire team and the length of the shift could itself be varied. The friendship associations of the team and the development of shared expectations about behaviour provided the source of a number of possibilities for feed-back onto the technical system, filtering its demands and producing distribution of load through the elaboration of certain modes of action.

The individual roles in the work crews, as well as being instrumental to the attainment of a given level of material output, also achieved sub-goals which increased the comfort of operatives: greater variety was introduced and a relief against the pressure imposed by a repetitive work cycle.

Thus far we have referred to aspects of work behaviour which were observed in the pre-automated technology. Figure 5 also suggests that there were shared attitudes which were connected directly with these. The interviews revealed a frequent and consistent expression of the values which were current among the factory employees and it was therefore possible to chart, in outline, the norms about which there was consensus.

Crew norms

Briefly, the most important norms touched on interpersonal relationships, work effort and leadership in the crews: defining the work performance expected of each man and his obligations towards others in his crew, they provided a general programme which was coded eventually into the different regulators. These statements of appropriate actions or responses which were expressed in the interviews specified the tolerances of behaviour for the individual towards others and also what he could expect from them.

If an order of importance could be assigned to the norms, then what the men referred to as 'understanding' or 'harmony' would be foremost in ranking. Where the work operations were so directly interdependent, a high premium was set on mutual help and reciprocal services, because of the immediate consequences of failure as expressed in the remark of a production operative that 'one man can pull a pack to pieces if he does not pull his weight'. The 'harmony' norm was illustrated in references by the men to relationships in the crews and the effect of these in reducing work pressures. For example, a second helper pointed out how 'we can help one another and we are ready to help one another, that is the feeling of the mill'.

A question on the relative importance of operative jobs, generally drew the answer 'every man is important in the mill'. With the recognition of the importance of each member went a requirement for his full contribution, carried in statements like the following: 'If you get one man doing nothing it makes it hard all round'. This expectation extended beyond performance of job operations to reciprocal obligations for punctuality and regularity in attendance at work. Crews agreed on the times of starting the shift, some starting earlier than others as arranged with the crew of the previous shift. Deviation from these agreed times met with sanctions from other members ranging from verbal disapproval to the resolutions passed in the union branch which are discussed later in the chapter.

It can safely be assumed that the shared attitudes among the crews were partly an effect of resonance,[2] that is where the area of interdependent tasks produced perceptions of similar problems and echoing of the individual's opinions or beliefs among others in his group. But there were also structural features which would tend to produce law-abiding, in the sense of uniform, behaviour in work groups: most evidently in the way that their composition was determined. This was done by a team choice mechanism which had been formulated among the production workers, and which allowed operatives to follow their own preferences in finding a crew.

Team choice

The self-selection procedure was operated through the production workers' union; requests for vacant places in crews or for transfers from one crew to another were dealt with at the local meetings. Following the announcement of vacancies in crews by union officials, members could request a promotional move to the vacant place or otherwise transfer from a similar position in another crew. These moves were formulated in the union records which were treated by content analysis of branch minute books from 1937 to 1956. I attended union meetings to observe the operation of these procedures during 1956 and 1957. The business of filling vacancies in the crews took up just under a third of all union activities in these branch records.[3] An item analysis of all the separate entries or minutes showed (as Figure 6 illustrates) that in some years this proportion could be even higher.

The procedure for self-selection of individuals to a given crew was as follows. When a vacancy in one of the operative grades was announced, it was open to any member already occupying the equivalent grade in another crew to claim priority in moving into the vacant position. This took place before members in the grade below, who were next in line of seniority, could be considered for promotion.

So it was possible that there might be a series of moves following the announcement of only one vacancy and involving several men in the same grade before the promotion sequence itself was started. The local union had stated this procedure, as follows: 'the senior hand in every grade should have the privilege of changing mills, if dissatisfied with the crew or conditions'. Similar branch resolutions also gave operatives the right of three consecutive moves between crews, as vacancies occurred, within any one of the production grades.

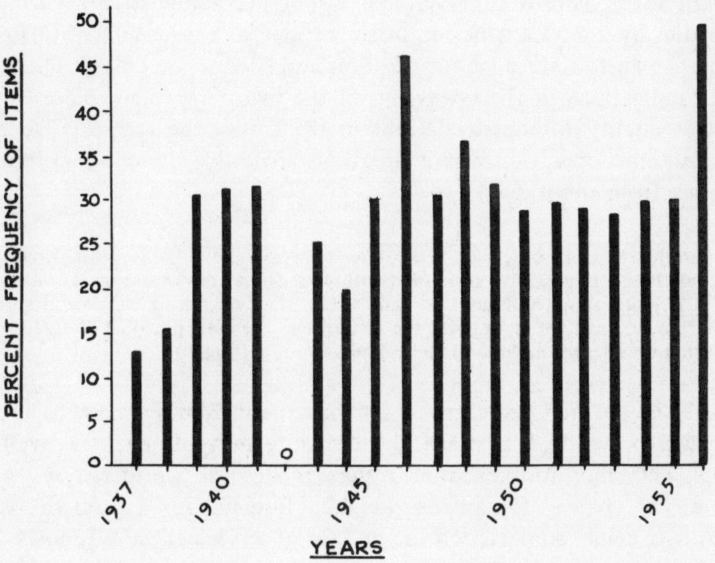

Figure 6. Frequency of 'Vacancies' 1937-1956 – Item Analysis

This mechanism was extremely well suited to the needs of the individual. For example, a member whose request for a move was granted could, if he wished, return later on to his original place should it not work out as expected. Crew changes arranged privately between members could also be formalized subsequently in the meetings: the members exchanging equivalent positions in their respective crews.

As mentioned above, the union rules allowed for two kinds of move. First, there were moves sought for technical reasons. Certain of the machine rolls worked heavier steel bars and produced higher average wages for their crews. The individual's choice here might be made to a mill with these heavier orders, in favour of increased earnings but with the disadvantage of greater physical load. Occasionally, there were instances of the reverse type of transfer where operatives expressed a preference for machines with a lighter throughput of bars. In the former case it would depend evidently on whether the positive incentive of higher earnings outweighed the disincentive of increase in load. In the case of a move to a lighter machine, this type of preference tended to occur with the changing need balance in older workers and the decline of their sensorimotor efficiency.

Secondly, a member might move to be with a friend or avoid an enemy. It was interesting here to observe the union branch administering work group composition by a device which in other settings, as a management technique, has been shown to reduce labour turnover and production costs.[4] There were repeated indications in the interviews of the high importance placed on the supportive quality of social relations in the crews; the team choice mechanism came into operation where situations arose like the one mentioned in this extract from an interview:

'To get any decent work out, every crew tries to work as one if they can. You have to have a good crew with you, five good men with you. Some crews seem to get on together better. With some crews, you get one who doesn't fancy working with them so much. Sometimes he just can't keep up with the rest, or else there are personal disagreements. I have had that myself, a doubler and second helper not getting on'.

Because of the freedom allowed for movement from one crew to another, the probability was high that any given work team would be attractive for its members. The union formalization of these moves was considered by the men to give a 'fair chance' for an individual to find himself a team to his own liking. A first helper summarized the process of self-selection as follows:

'You would always look for a pack that can get the work done, one that's a team. A furnaceman that understands his work and a rollerman every bit as particular as the furnaceman. And most important, you always look for a pack that respects each other, if this is absent then without deliberately trying, a man becomes haphazard and the job is more hard than it should really be'.

In these two extracts, it should be noticed that interviewees referred to the goals or standards of work performance to be achieved as well as evaluating the crew in terms of its degree of interpersonal agreement.

Composition of the production crews

We now see that the crews were likely to be attractive to their members because of the mobility which was possible for individuals and the cumulative operation of members' choices. The composition of the crew tended to be formed through the expression of friendship ties between members or, at a minimum, to be built up on the similar working capacities of members. This basis of crew formation would in turn make it even more likely, than the incidence of what was earlier referred to as resonance suggests, that there would be consensus on the sequence or pace of work routines and the norms to be followed in this respect. In every team a high degree of coordination could be expected and, as supervisory rating confirmed, this was so for most of the crews. It was estimated from such ratings obtained for all crews and confirmed by the interviews with rollermen that only three out of twenty teams had failed to achieve this degree of consensus.

It seems likely that the realization of individual choices would give membership of a particular crew a high value and there was some additional evidence of the attractiveness of the work group. For example, it was usual to find the group-centred perception among interviewees which some studies have connected with high performance levels.[5] The belief that one's own crew was better than others was expressed in such terms as the following, where rollermen give their opinions of their crews:

'You may think I am being very proud but I had the best furnaceman in the mills claiming to go with me when I became a rollerman. It made me very happy'.

'The gang I've got, I'm not boasting—but you could ask them—I wouldn't like to force anybody (to work faster) and I'm quite happy with the gang here and I'm sure they feel the same. You can look at the report of mill output, you wouldn't have to look further for what sort of crew I have'.

The promotion procedures

To summarize, the team choice mechanism and promotions into each of the grades were dealt with as a routine procedure by branch officials, from lists of names drawn up by the branch committee. It was also open to members on their own initiative to claim vacancies. While it often happened that members presented their claim for promotion at the meeting, the system was also flexible enough to allow members to refuse a promotional move if they preferred a job with lighter duties.

In most cases the procedure carried through smoothly with the longer service operatives in any one grade moving to vacancies in a grade above. The hierarchical structure of work teams consisted of a promotional ladder or line, with eight grades distinguished by title, duties, wage-level, and by skill or expertise (see Chapter 2). There were six crew grades and two additional positions: bundler, the grade entered by school-leavers on arrival in the factory and shearer, the culminating position of the ladder, whose tasks involved preparation of sheets for dispatch to the next production stage.

The order of succession is represented in Table 1.

The promotional line could be followed through each of the crew grades to rollerman as in Table 1(a). In practice the line tended to be the one shown in Table 1(b). Moreover the second helper grade would often be missed out in the advance upwards because of the close similarity in duties of the two bottom grades.

Table 1. Steel-rolling crew, promotional lines

(a)	Shearer	*(b)*	*(c)*
Rollerman		Rollerman	Furnaceman
Doubler		Doubler	First helper
Furnaceman		First helper	Second helper
First helper		Second helper	Back-roller
Second helper		Back-roller	
Back-roller			
	Bundler		

The rollerman grade could only be reached from the grade immediately below (doubler), which was the most difficult of all in terms of muscular exertion and physical load.[6] For this reason many operatives preferred to progress, as in Table 1(c), only to furnaceman which although a senior position was less physically demanding. The final possibility in promotion (with the exception of appointment to the supervisory positions) was the grade of shearer, a position with large responsibilities for quality and the best paid in the factory. Translated into individual terms, the promotional line provided an assured progression over fifteen to twenty years or more.

Skill hierarchy

With six positions in the work team each differentiated from the other by title, duties and wage-level, a steady advance along a promotional ladder was guaranteed to the steel workers; illustrated in the experience of this production worker:

> 'I was back-rolling for 2 years, then second helper for 12 months, doubler for 12 years and rolling 16 or 18 years up to now'.

It appeared that this promotional advance through the mill grades was not perceived as slow in a work system where the skills and responsibility of senior grades were fully recognized; each operative, providing he could meet the skill requirements, could be fairly certain of promotion to the top positions in the team. Extending over a period of years, the skill hierarchy was formative for the individual in affording him advancement and increasing recognition for his accumulated skills and experience.

There is no doubt that the promotional allocation was the most basic of the rewards available to a member from his own attendance at branch meetings. At the same time, promotion was also a major reward available to the individual from the primary group for his attainment and conformity to group norms. The union requirement for promotion was seniority 'and efficiency', and the latter tended to be assessed in terms of the norms both for work performance and for interpersonal behaviour. Consequently, there were examples in the minutes of branch members exercising a bar on the promotion of an individual by reason of his deviation from either.

Demotion and semi-retirement

The promotional line could operate in a downward direction (demotion) for those employees who wanted to withdraw from the heavier load and responsibility of the senior grades. The second helper grade afforded the most easy 'semi-retirement' position for the older operatives. This was entirely a matter of the individual's felt need and took place only on the introduction of the member's request during the meeting. Work-load strains on older operatives were thus modified by a procedure which was ideally suited to individual differences in the ageing process. This type of move was similar to transfers between crews using heavier rolls to those on lighter machines, which were described as following a change in need balance with advancing age.

Interpersonal conflict

One of the consequences of union control over promotion and team choice were the disputes which arose from members' claims for vacancies or promotional moves. Figure 7 shows how these 'seniority disputes' might occupy as much as fifteen per cent. of branch business.

The usual procedures for dealing with contested seniority were as follows: when more than one member claimed the same vacancy or promotional move they would be asked by the chairman to state their claim to the meeting. In order to win, the member was required to produce evidence of longer membership. But his position might not always be easy to establish, for example, where dues had lapsed during absence from work, or again where members had joined the company within a few weeks or days of each other. It was usually the case, where precedence could not be established during a meeting, that the dispute would be referred to the branch committee and in the meantime the officials would consult union records for information which could resolve the dispute.

On the face of it, the presentation of a promotional case might have been no more than a matter of informing the meeting of a set of facts. However, the advocacy of claims often became heated verbal contests. The branch minutes refer to a 'warm discussion' or a 'bitter controversy' and give an indication of the antagonisms characteristic of these disputes between members with the officals acting as mediators.

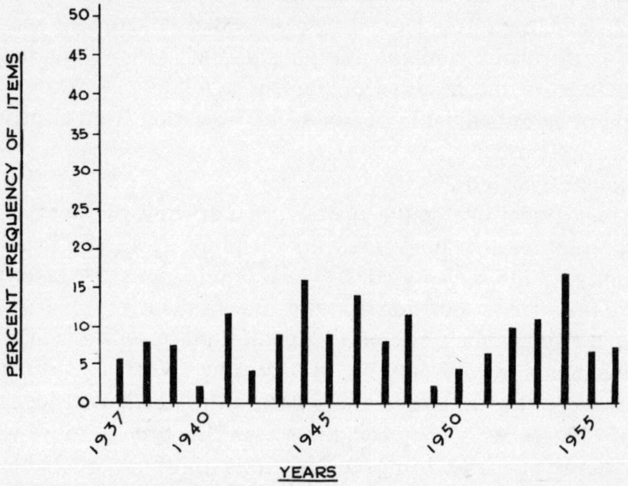

Figure 7. Frequency of 'Seniority Disputes' 1937-1956 – Item Analysis

The expression of interpersonal conflict on these occasions appeared to go a good way further than simply a statement of claim and counterclaim. The interpretation is suggested of the branch providing an outlet for combativeness which would have, in the mill crews, hampered the cycle of operations. A further interpretation could be that the seniority disputes took the place of selection procedures–notions of distributive reward in the primary group being satisfied by this particular form of seeing justice done. This in itself would be likely to remove sources of friction in the social relations of the work crew.

As a locus for the expression of personal differences and at the same time a means of containing them the branch thus protected the smooth functioning of primary group relationships in work. Seniority disputes took a form similar to the Scottish 'flytings'–the institutionalized exchanges of verbal abuse, where contestants exchanged insults and ribaldries before an audience. It is interesting that during seniority disputes, although the displays of interpersonal aggression could be intense, the officials were able to restrain such outbursts effectively, so that afterwards the minutes could still record: 'both disputants shook hands and left the room the best of friends'. By leading to resolution of conflict and tensions between individuals the meetings appear to follow the problem-solving sequence which Bales derived from his analysis of behaviour in small groups.[7]

In the section on behavioural sanctions it is described how the branch acted in yet more direct ways to maintain the system of interpersonal relations which existed in the rolling mills.

Behavioural sanctions and the union

In a sense the branch was acting as legislature for the social group in formulating procedures for promotion and transfer and as a judiciary in settling individual claims for vacancies. Yet another class of branch activities lay in implementing or enforcing the behavioural norms of the work shop. This was action of a comparatively rare type because the rewards and sanctions (for conformity or deviance) were usually administered through interpersonal communication in the work team. Nevertheless, in this situation where operatives met at the local union branch over a long period, boundaries between formal (branch activities) and informal (social group) events became less sharply defined insofar as meetings could become an extension of primary group interaction and a medium for group processes.

Let us consider some examples. In relation to work performance the norm was mentioned earlier of maximum contribution of effort on the part of each crew member. Sanctions appeared in the minutes in the form of complaints that certain named individuals were 'not doing the work to the satisfaction of the rollerman' or 'not pulling their weight'. Crew members would express their refusal to work alongside them and displacement to the lower paid job of relief worker might follow. Examples of the most severe sanction–requesting a member's dismissal–were only found in the case of extreme deviance by immigrant workers. Both the introduction of workers from Eastern Europe in the late 1940's and from Italy in the 1950's produced an increase in branch actions as an agency for sanctions, which Figure 8 illustrates.

Another example of implementation of crew norms arose where the branch mediated between individuals, i.e. to settle disagreements occurring within a crew during work. The importance of cooperation in the teams has been shown in the controls which were possible over pressures of the repetitive job cycle. Obviously, disagreements between two individuals could make work tasks generally more difficult and in some cases recorded the members concerned were reprimanded through the branch meetings.

As with all norms, conformity was a matter of degree–as we have seen, a few crews were known not to have achieved smooth relationships and it could be inferred from the interviews that they suffered loss of social esteem and low ranking in consequence. These were the same crews that were rated lowest by the supervisory staff in relation both to interpersonal behaviour and work performance. But with the transfer mechanisms which existed it was more than likely that most mill crews would be well matched in their composition.

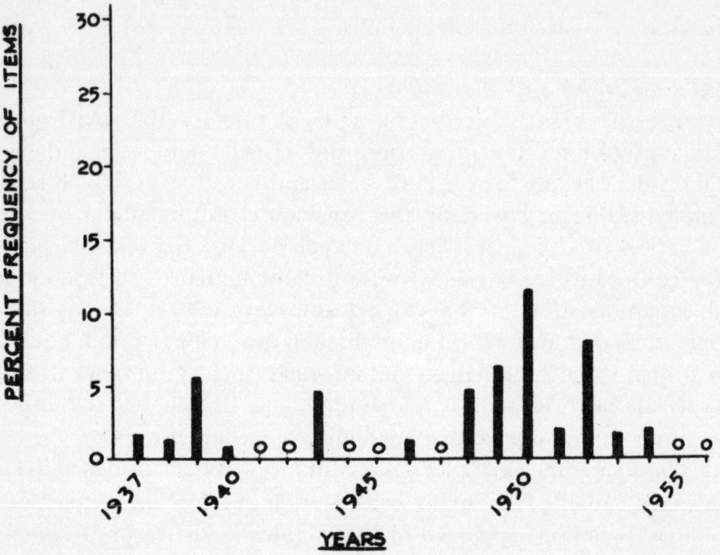

Figure 8. Frequency of 'Behavioural Sanctions' 1937-1956 – Item Analysis

The structure and composition of production crews; the rewards which were directly available from collusive activities; the behaviour which has been analysed in this and the preceding chapter, need to be borne in mind for the subsequent comparison with the team in the new technology. The social induction and informal training of new entrants to the work shop and the opportunities for steady promotional advance were associated with closed-loop behaviour achieving a number of sub-goals. The team choice procedures were implemented by the local union and it was always possible to escape from a less congenial team. Each primary group had attraction for its own members, and in these conditons it evolved social regulators which allowed the crews to operate more effectively, acting on the heavier job demands to modify them in the individual's interest. When the men were transferred, would the pattern continue, disappear or reformulate? The consequences in the individual will be of special interest in the description of the innovation effects.

Summary

As a result of behaviour arrived at by consensus among the production personnel, the individual received a number of rewards in the form of additional relief breaks and increased variety through exchange of his tasks with others. Furthermore, collusive action in the team led to changes in the pace, sequence or duration of the work cycle. These forms of behaviour operated as social regulators on the demands of the technical environment, in the closed-loop manner which was illustrated in Figure 5.

It was interesting that the local union branch had control over promotion procedures and the filling of vacancies. In this and a number of other instances which were described the union branch intervened in the workshop to influence the composition and objectives of the production teams. The way in which seniority disputes were dealt with by the union appeared to be a form of selection procedure as well as a means of removing sources of friction in the work crew.

REFERENCES

1. The role of the union branch in applying sanctions against individuals is more fully described in J. K. Chadwick-Jones, 'Acceptance and Socialisation of Immigrant Workers in the Steel Industry', *Sociological Review,* **12**, No. 2 (1964); and by the same author 'Italian Workers in a British Factory', *Race,* **6**, No. 3 (1965).
2. The concept of resonance is defined by W. F. Whyte in 'An Interaction Approach to the Theory of Organization', in Mason Haire (Ed.), *Modern Organisational Theory,* John Wiley, New York, 1965, p. 162.
3. The analysis of the subject-matter of these union branch records is presented in detail in J. K. Chadwick-Jones, 'The Union Branch: a Socio-psychological Interpretation', *Occupational Psychology,* **39** (1965).
4. H. Van Zelst, 'Sociometrically Selected Worker Teams Increase Production', *Personnel Psychology,* **5** (1952).
5. S. E. Seashore, *Group Cohesiveness in the Industrial Work Group,* Ann Arbor, Michigan, Survey Research Center, University of Michigan, 1954.
6. Attention is drawn to the heavy physical load on the doubler in the early reports made by the *Medical Research Council* and *Industrial Fatigue Research Board* on the steel industry.
7. R. F. Bales and F. L. Strodtbeck, 'Phases in Group Problem-Solving', *Journal of Abnormal and Social Psychology,* **46** (1951).

Chapter 4

Factory and Community: Before Innovation

An examination of the community context both of the new and old industrial plants will be useful because it touches on the interpretation of the men's attitudes towards their jobs and on the understanding of work group, supervisory and managerial relationships in the firm. A description will be given of some of the influences of community and leisure group on work behaviour, as they were revealed by the interviewees' own perceptions and expressions of opinion. This chapter therefore allows an insight into the ways in which community and factory were interrelated. Once this relationship is spelled out then we can, in a later chapter, examine how it was affected by the technological innovation. After the changeover there was some direct collection of data in the community by interviewing wives of the steel workers. Although the community is not a central area of the analysis, aspects will be included which were of significance for the individual in his work group.[1]

The community background

The old factory was situated a quarter of a mile from the main traffic intersection of a township with a population of around ten thousand, where the steel industry had provided the major employment opportunities since the mid-nineteenth century. The town grew up around the steel and tinplate factories in the typical pattern of nineteenth-century domestic buildings– 'workers' houses, often those of the middle class too, would be built smack up against a steel works'.[2] The great majority of those interviewed (e.g. eighty-six per cent. of the production workers in the hand mills) were born within a mile, the remainder up to three miles from the factory. This community, as could be expected, possessed a separate and distinctive identity of which there was a strong awareness among the steel workers, in relation to the surrounding urban centres with more varied occupational strata. It was similar to the communities described by Rosser and Harris[3] and Lockwood[4] as strongholds of a traditional occupational culture associated with a period of stable residence, spanning several generations of workers in the one main industry of the region.

This is important for our analysis because of the influences on the attitudes and behavioural norms held inside the factory which came from community sources. It also assumes importance for understanding the implications of the transformation in the steel industry. The existing community structure, it will be demonstrated, had relevance for the acceptance or rejection of certain effects of innovation by the individual.

The importance of kinship connections in a long established community with little population movement was expressed in the local saying that 'our town is one family', and as might be expected there was a network of friendship and acquaintanceship developing from proximity of residence and common membership of churches and local societies. In the town's one main street it was always possible to meet steel workers. At the end of a shift, groups tended to form at the town centre where the probability of conversation turning to topics related to work was expressed in another current saying that 'more steel is produced on the town square than in any steel works that's ever been'.

The 'family spirit' of the factory which was often mentioned in the interviews reflected in part the nature of this community and its values. It has already been shown how these values were implemented in work group roles and how they affected relationships with supervisors and management. Given the direct role of the labour force in the production system, and a payment system by group piece rate, there were frequent occasions for plant bargaining and management-worker disputes which were argued out between top management and deputations of union officials. However, the fierce opposition recorded in the minutes of the union branch has to be understood as happening against a network of kinship and neighbourhood connections throughout the organizational hierarchy, and the existence of such outside ties made it less probable that disputes could be taken as far as strike action. Any strikes which had occurred (and there were very few in the history of the firm) were the result of industry-wide movements, not of local disagreements.

Community ties

We shall take the description of the community and its relationship to the factory as far as the interviewees' perceptions lead us. Information is available on the community life broadly of this region in the study by Rosser and Harris,[5] which elaborates in great detail what was here indicated by the managers, supervisors and workers in the factory concerning their leisure group memberships and cultural or class allegiances. The interpenetration of community and work norms will become very evident as we go on.

The powerful influence of the local neighbourhood was shown by the extent to which it exclusively supplied the labour force to the factory from one generation to another. Older employees narrated how, at an earlier period, entry to the firm was restricted and, if not through a near relative, being 'spoken for' was at least necessary through neighbourhood friends. One man described how he and his brothers were 'favoured' by one of the local entrepreneurial families which owned the factory:

'I have been here since 1922. We were four boys, my father was in the works and my elder brother was with him. There was a bit of preference on the Johnsons' part in getting us in'.

Paternal or other family relationships appeared as the traditional preconditions for admission into the old plant, the production jobs tending to become a pseudo-property right inherited through relatives: this type of entry filter and the close bonds between neighbourhood residence and the factory were similar to those of the dockers' community described by Young and Willmott.[6] Entry to the steel plant was by father-son succession in the case of more than half the interviewees of both samples. Forms of succession by kinship were maintained up to the time when the technical obsolescence of the plant and cessation of recruitment undermined the existing social connection between plant and neighbourhood.

Certainly these conditions had existed for a period of at least several decades before the changeover. But during the ultimate obsolescence of the small steel plants this state of things and the nature of the community itself had been changing into more diverse forms than when a single industry dominated its immediate local community. Under these former circumstances the expectations of the local workers' children had focussed on their future employment in the steel mills. Brought up 'within the shadow of the works', they were encouraged to aspire to the occupations of their father or uncles in the grade ladder of production jobs. Introduction to the work and conditions of the industry began very early with the children carrying freshly made meals to the rolling crews from houses just a few yards away. The following extracts from interviews show how effective was this kind of induction. The first interviewee was a rollerman, the second a supervisor:

'My father was a rollerman. I used to take dinner to my father in the works. My brother was also a rollerman and my sisters worked there'.

'The steel works is in my blood. I've been brought up in that atmosphere and I've always been interested, very, very interested, and since I was a boy I have always been ambitious. The goal I had as a boy was to be superintendent of a mill. I've struggled through every grade in the mill to get on. As kids I remember playing at mill work. We used to come in on a Saturday and pretend to pass the piece through the rolls. In the house we used to pass a piece of wood from one to the other'.

These men conveyed the atmosphere of their boyhood with clarity, and they conveyed also the comprehensive extent of their personal investment in the industry. The degree of their commitment to a certain type of work forced an awareness of factors in relation to which the effects of innovation on the individual and his reactions to it might be more adequately understood. One worker, asked after the transfers to the new plant 'what is the main difference between working here and your old job?' replied, 'the change in my life': it was the identification of the job with a *basic life interest* which was so evident among these interviewees. In face of this, conventional ideas for job re-training might imply too superficial an approach to the individual's problems.

Kinship among the labour force

The prevalence of kinship was already diminished at the time of the interviews by the tailing off in recruitment from the local population for a period of some years before this study began. Because of the limited future of this section of the industry, no local workers were engaged and temporary labour shortages were met by unskilled workers brought from Italy on short-term contracts. Since kinship ties were therefore not being renewed by the entry of younger members of the production workers' families, kinship at the time of the interviews showed a corresponding difference in the number of present ties as compared with past connections. Full records were available for twenty-one members of the sample and these had forty-seven past connections (mainly grandfather, father or uncle) but only seven present ties (brother or cousin). None had sons in the old factory.

The average of both past and present connections combined among these production workers before the changeover was less than half the average recorded for production crews in those sections of the steel industry not under threat of immediate closure.[7] However, the kinship connections did indicate homogeneity of the labour force and it would not be surprising if characteristic forms of behaviour had evolved over time.

Community influence on management

Another illustration of community influences on work behaviour derived from neighbourhood friendships. There were instances where the wives of workers had intervened with managers' wives to produce a relaxation of rules. In one interview a supervisor described how his wife had been chided by the wife of a production worker concerning disciplinary decisions made by him. A departmental manager recalled how he had told three men that they were suspended from their jobs– only to have them reinstated after their wives had talked to the wives of more senior managers.

The social structure of the factory, influenced as it was by the community environment, itself extended into community activities. Employees tended to meet outside of work, thirty-nine per cent. of the rolling mills sample saying that they met regularly. One rollerman mentioned meeting at a nearby town to attend football games:

'The bar cutter, two doublers, the furnaceman and the first helper, we're always together in football. I meet my furnaceman in town and the first helper and the two doublers also go the match every Saturday with us'.

Church membership was another focus for meeting, especially for the wives of both workers and managers, and was an additional source of wives' friendships. The senior production workers tended to hold lay offices in local churches: of the eight men at the apex of the grade hierarchy—the shearer grade—six were in the highest lay office of local nonconformist churches. In one church the works manager and one of the shearers held similar positions, and in another the plant superintendent and a shearer were lay officials.

Within the management hierarchy there were a number of kinship and personal ties but between the managers and the hourly paid workers there was also a wide range of kinship. These relationships were varied and extensive; just how pervasive they were was easy to realise when one was told that the commercial manager's brother was a labourer. Or when a second helper told me that 'my cousin is the managing director's son-in-law'. In the rolling mills the superintendent had a brother and brother-in-law in the production crews.

Supervisors were always addressed in first name terms by their subordinates, who shared an identical class, cultural, and work background: 'It's as if he's one of us' represented frequent references to this common social identity. The network of personal relationships which has been outlined involved the management of the firm in loyalties or obligations which tended to induce behaviour towards employees following alternative lines of action to those required by the rational discipline of business efficiency. This revealed itself in various ways, such as the care taken of old employees who were found light jobs in the company. Some of these were over seventy. The most extreme example was the employee who was seventy-eight with sixty-five years' service. The general manager was the son of a former supervisor in the factory and, so it was said, started work as a boy 'sweeping the floor of the bar mill'. Several of the employees recalled how they were in the same class in a local elementary school or how they were taught at school by his brother. Some of the managers were not local men and they were unable to participate to the same extent in the social close-weave of the factory: for them there were 'too many deacons and cousins, and people who made cups of tea for the general manager forty years ago'.

In the course of the changeover to the new plant the community influence was dispersed by recruitment from a wider urban region. How this affected workshop behaviour is described in the later chapters. The effects of factory innovation on community and leisure activities–especially those of the new shift arrangements–will also be discussed.

Summary

The influence of the immediate neighbourhood was very strong in the period before the new plant was commissioned. Most interviewees were born within a mile of the old factory and the production jobs tended to be filled by relatives and neighbourhood friends. In these circumstances induction to the factory began very early and the men described how their aspirations to become production workers were formed. Neighbourhood associations and kinship affected both the relationships between managers and workers and management-union bargaining. Managers, supervisors and senior production workers shared leadership positions in local societies and churches. While the influence of family and neighbourhood was carried over to the factory, work affiliations were also extended into leisure time.

REFERENCES

1. Tom Lupton discusses the problem of analysing influences from the social environment external to the firm on 'in-plant' behaviour in *On the Shop Floor*, Pergamon Press, Oxford, 1963.
2. Lewis Mumford, *The Culture of Cities*, Secker & Warburg, London, 1946, p. 163.
3. Colin Rosser and Christopher Harris, *The Family and Social Change*, Routledge & Kegan Paul, London, 1965.
4. David Lockwood, 'Source of Variations in Working Class Images of Society', *Sociological Review*, 14, No. 2 (1966).
5. Colin Rosser and Christopher Harris, *The Family and Social Change*, Routledge & Kegan Paul, London, 1965.
6. Michael Young and Peter Willmott, *Family and Kinship in East London*, Routledge & Kegan Paul, London, 1957, p. 75. Here the father-son succession among dock workers is described, the job becoming in effect a 'pseudo-property right'.
7. W. H. Scott, J. A. Banks, A. H. Halsey, T. Lupton, *Technical Change and Industrial Relations*, Liverpool University Press, 1965, pp. 51-52.

Chapter 5

Job Satisfaction in the Old Plant: The Anticipation of the Changeover

We will now examine the positive or negative features of their work as they were perceived by the men in the pre-automated technology. Overall attitudes were classified from the replies to a direct question *How well do you like your job?* into three broad categories: satisfied, dissatisfied or undecided. Scaling techniques were not used on these replies; it was thought more important to allow the interviewee to speak freely and to encourage an easy conversation which presentation of elaborate check-lists or ratings would have made difficult. The main emphasis of the procedure rested on giving the men confidence to express their opinions, with the aim of drawing out information especially on details of context which it would not be possible to cover by questionnaire methods. It was therefore decided to surrender certain quantification possibilities for the sake of gaining access to topics and even minutiae of behaviour which could not otherwise be revealed.

Constraints to acquiesce in the situation

It is a well established finding of many industrial attitude surveys that the majority of employees express satisfaction in some degree, if only by the sheer fact of continuing occupancy of the job and longer occupancy itself tends to be associated with increased satisfaction. We can assume, however, that the degree of liking may tend to be lower among those who find themselves in a given occupation by constraint rather than by individual choice. Under ideal conditions an individual would be able to enter an occupation because of its greater appeal or attraction among a number of alternatives. It is clear from research studies that the limitations on individual choice are greater for some occupational strata than others: for lower social-economic categories possibly more than for professional levels and, within the hourly-paid categories, more for those living in traditional one-industry regions than for those who live in highly diversified urban developments.[1] As was shown in the last chapter, the steel workers did not choose their jobs out of a range of alternatives; they were, as one rollerman said, 'born to it' together with most of their contemporaries in that community. But we also saw that although there were constraints of this kind, socialization through family and neighbourhood meant that the individual's own aspirations were likely to be directed towards the industry.

We could hypothesize that constraint pressures were much greater on the interviewees after the innovation had dictated their transfer to the new technology. At least in their previous jobs they exercised some, albeit restricted, degree of choice in entering the industry or section of the industry (melting shop, barmill, tinplate and so on) in the first place; now after years of employment the only alternative to the transfers was the somewhat unlikely one of leaving the industry altogether. We could therefore consider that a man's expression of general satisfaction, especially after the transfers, might well reveal the extent of his *acceptance* of the job rather than its *attraction* in itself. In the light of these pressures to accept a given situation, statements of general 'satisfaction' with the new jobs might reflect, more than anything else, the awareness on the part of the employees of its inevitability.

The analysis in detail of specific likes and dislikes both before and after the changeover will treat with the difficulties or improvements encountered, the gains or losses from the innovation. In this chapter the attitudes expressed towards different aspects of the job and its environment are discussed for what they show of the rewards or otherwise of the old jobs and the anticipation, positive or negative, of the approaching transfers. Later on in the book, the question of how far job satisfaction may really mean acquiescence will be discussed more fully.

Anticipatory effects of the changeover

The overall response tendency towards the jobs in the old factory showed the following distribution of replies:

Table 2. Overall attitudes towards the job (*N* = 47)

Satisfied		*Dissatisfied*		*Undecided*	
No.	*Per cent*	*No.*	*Per cent.*	*No.*	*Per cent.*
33	70	9	19	5	11

Many of the sample had long service in the industry in occupations with skills which were easily identifiable and prestigeful in the community, and for these reasons the distribution reveals a heavier negative bias than might be expected among such a population. The explanation was found in the special circumstances of the approaching innovation–which had been publicly announced and was known to be imminent. The effects on the employees thus extended in advance of the changeover itself: their jobs could have no place in the new extensions of the industry and dissatisfaction arose from the consequent lowered evaluation of their occupation. There was also another factor–most of the employees had not been told whether or not they would be re-engaged after the closure. Almost a third of the sample expressed outright dissatisfaction or had become undecided on whether they liked the job or not. The latter expressed minimal satisfaction as implied in the comment 'it's a job, I've *got* to like it', qualifying such statements with further unfavourable opinions.

The analysis of likes and dislikes relating to specific job or work topics before the changeover will now be described. The results are tabulated in Table 3, below.

The free-answer data

References to specific features of the job were made in *free* answers to general questions which left to the interviewee initiative to mention topics in his immediate or peripheral work environment. Analysis of these allowed a check on the overall satisfied, dissatisfied, or undecided categories by a comparison of the ratios of likes to dislikes expressed by interviewees in each category. The use of ratios also made it possible to check on the equivalence of the categories of satisfied, dissatisfied, etc., established for this first sample with those for the interviews after the changeover.[2] In all, three measures have been used in the analysis of job satisfaction in this chapter and later in Chapters 7 and 8. They are derived in the following ways:

(a) By taking the overall numbers of satisfied or dissatisfied employees before and after the changeover.
(b) By listing the topics mentioned as likes or dislikes in rank order by the number and percentage of those mentioning them. This provides an order of importance of the various factors as perceived by the interviewees.
(c) By summating the favourable and unfavourable opinions for each group and then calculating ratios of favourable to unfavourable comment.

We can now consider the references to specific positive or negative characteristics of the old jobs, which are shown in rank order in Table 3. Before discussing each item it is important to be aware of the underlying assumptions about this method of data collection.

Table 3. Job satisfaction in the old Plant (*N* = 47)

Likes	*No.*	*Per Cent*	*Dislikes*	*No.*	*Per Cent*
Familiarity	16	34	Physical work load	20	43
Environmental conditions	15	32	Inefficiency of plant	12	26
Intrinsic job factors	12	26	Obsolescence factors; loss of pride in work; lowered status of jobs	9	19
Social relations	10	21	Wages	3	6
Hours, shifts	10	21	Leisure curtailment (through fatigue)	2	4
Management fairness in wage agreements	7	15	Inadequate factory welfare provisions	2	4
Supervision	2	4			
Freedom in work	2	4			
Wages	2	4			

The tracer effect

There is what might be called a tracer effect in the distribution or weight of unsolicited responses. In other words, if a topic or job feature is shown to be important to a given frequency by this method it can be assumed that, were a direct question to be asked to all members of the sample, the response tendency indicated by the free-answer method would be proportionately augmented. The percentages in Table 3 would not be considered high if they had been yielded by a coverage of each topic by direct question, but in this case since they represent spontaneous comment their meaning is quite different. Clearly the method will underplay the latent response tendencies of the sample because there is no prompting or persuasion on the interviewee to reply on any given topic. Therefore, while those for whom a job characteristic is sufficiently important will refer to it, there may be others from whom it has value but not to the extent of expressing an unprompted opinion. The topics which are foremost in the rank order in Table 3 tended to be mentioned by thirty to forty per cent. of the interviewees, and the particular conditions of this response rate need to be remembered in attaching due importance to these relatively (and apparently) low percentages.

It is necessary to take into account also the limits on the articulateness of the interviewees. Some of the rollermen could provide as many as eight topics or job characteristics to which they put positive or negative valence—on the other hand, two of the back-rollers could each give only one job aspect. After the changeover the number of topics which were mentioned showed an increase, presumably because the interviewees tended to relate their new jobs to the old, referring to the characteristics or rewards of each.

We need to examine the configuration of likes and dislikes in Table 3 as our basis for comparison with attitudes after the changeover. The new plant was in the final stages of its construction, a few of the men had already been selected for transfer but most were waiting to be notified. The immediate future was still uncertain; it was known that the plant could run for a limited period but no-one had received information on how long it might continue. The unknown consequences to each individual produced conditions for the spread of rumour and the likelihood of imminent closures was frequently discussed in everyday conversations. To some extent both positive and negative attitudes to the job were affected. The men's dislikes showed a preoccupation with inefficiency, obsolescence and the physically difficult operations of the hand mills which they contrasted to the likely benefits of innovation; and they were aware of the lowered status of their jobs in the community in general. Among the positive features mentioned, the references to the familiarity of their workplace and its jobs, almost a clinging to the familiar, might be connected especially to the situation of uncertainty.

Job satisfactions before the changeover: Familiarity

First in rank order by the frequency with which the topic was raised is the familiarity of the job. This topic could be sub-divided into two kinds of response, both of which appeared to develop over a period of long-term occupancy of a job. Primarily there was attainment by the individual of a condition of physical adaptation: 'it's a very hard job but I don't find it too hard because I've been used to it'. Habituation to the job accounted for more than half of the answers classified under this heading, the remainder reflecting a more general influence of familiarity with the workplace, a certain system of discipline and a group of work mates. Familiarity in this latter sense had its origins even in the induction of new entrants from boyhood: 'I like the job because I've always been brought up to it', a furnaceman said, 'your heart is here somehow and you don't feel like getting another job'. These were attitudes which may be expected where workers have known no other type of employment. 'I've always worked here and would find it strange working anywhere else' was a typical comment in this category. On the other hand, those employees who had had experience of different jobs, usually for temporary periods during wartime, explained they had discovered they 'couldn't feel at home' in other employment: 'Although the work is hard there is something about this place that seems to grip you. That is the funny part about it, although I have children and wouldn't allow them in. It's the no-discipline, you haven't got the foreman on top of you'.

Such attitudes should not be underestimated in situations of innovation. The analogy used by the interviewees to express their feelings is especially interesting as in the following reference to the plant as 'home' by another production worker: 'Your heart is here, it's like that you have two houses, you've made the works your home'. The extent of attachment to job and factory can, of course, be appreciated more fully when it is referred to the context of the social factors described in the previous chapter. Again, in the interviews after the transfers, when the men were trying to express the initial difficulties they had experienced in the move, some talked about leaving their 'home' for a new one. 'It's like moving, leaving your neighbourhood and going to an estate where you know nobody until you get to know them'.

Environmental Conditions

The emergence of environmental conditions among the positive features was an interesting outcome, particularly since the inefficiency of the plant was also mentioned among the dissatisfactions with the workplace. However, if the plant was inefficient in a general industrial sense, yet within that part of the steel industry which used the same methods and techniques, the pre-automated plant was better (in headroom, layout, ventilation) than other equivalent factories, nine of these being mentioned by name in the course of comparisons which the interviewees made. The following extracts from interviews are representative of these comparisons:

> 'I've seen other steel works and I think this is the best for space and place to work in . . . this is the finest mills I've ever been in. Better floor and better conditions than in others . . . it's one of the best mills in the trade undoubtedly. It's the best of all and I have varied experience of other mills'.

It is also possible that plant or group loyalties were increased in face of the threats of an uncertain industrial future. Favourable attitudes in this respect probably reflected both the hope and in some cases a desperate optimism that the old plant might continue in operation. Even so, when a question was asked on *how long the plant might continue working* almost twice as many were uncertain or pessimistic (seeing survival limited to a few months) as were optimistic (estimating the future as extending up to ten years).

Intrinsic Job Characteristics

Aspects of job tasks were mentioned by twelve of the sample: some of these thought the job was healthy and active, while five of those in the senior operative grades, rollermen or furnacemen, referred to the interest and responsibility of their jobs. One furnaceman gave a description of eventualities when sucn responsibility was needed, 'I like the job very much. There's more responsibility, if the sheets are scaling or are too hard, the furnaceman they come to'. The need to explain the task content and to describe eventualities or instances possibly accounts for only three referring to skill factors. At other points of the interview, for example, in answers to the question on the *relative importance of different jobs*, some of the men talked at length about the art of the mill worker. A rollerman pointed out how 'a pair of rolls have got to be nursed and coaxed exactly as if it's a child', and a furnaceman described how the steel bars had continually to be moved around the interior of the furnaces to maintain them at an even temperature for the rolling operations:

'You know where your eggs are. The furnaceman's art was taught me by an old furnaceman'.

Another reason why more did not discuss this interest and art in the context of *satisfaction* with their jobs is possibly supplied in the form of effects of the approaching transfers. The imminence of the closure of the old factory may have tended to reduce satisfaction in relation to task content especially since these skills would not be needed in the new plant and the jobs themselves would disappear.

Hours, shifts

Hours and shift arrangements were clearly to the liking of those employed. This was perhaps to be expected with a working day which might be of only six hours and on piece-work this period could be reduced even more. There were no formal clocking on procedures and employees could leave the plant when they pleased under the 'job and finish' system once their work quota was completed.

Six-hour shifts were being worked at the time of the interviews, this system had been introduced about a year earlier. Whether the plant operated on six or eight-hour shifts depended on market conditions and there had been several such switches in its very recent history. If sales conditions were favourable the eight-hour system was used, allowing a greater number of machines to be utilized with a lower per capita ratio of employees to plant equipment. On the six-hour arrangement a machine would be manned over twenty-four hours by four production crews—on eight hours by three. In the former, the men had to bear a drop in weekly wages; although this was a matter for complaint by only three interviewees. Some of the younger men occasionally undertook a double shift of twelve hours, for the sake of the wage increase. Otherwise, the advantage to the individual of the change to six hours was mainly in reducing a work load which often necessitated complete rest on arriving home after the shift:

'The only thing we are fit for after work is bed for an hour or two, to conserve your energy for the next shift. I take my leisure hours as quietly as I can'.

In addition to favourable references to the thirty-hour week under the six-hour shift arrangement, several of the men drew attention to the value which they attached to their free weekend in this industry. One interviewee considered his job was 'all right because I like the weekends, I like gardening. On any other job you've got to work at the weekend to make up your money'.

Social relations

The social relationships which were mentioned as a positive feature of the work were essentially *functional* as they were perceived by the interviewees. They were seen as varying the work, lightening the work load, making the job easier. There was no tendency to draw attention only to some general area of social satisfactions which was difficult to specify or, for example, to general conversational contacts. The references were very clearly to *particular* benefits for the individual from the supportive actions of the team. Mention was made of the attainment of consensus on shared roles: in all, twenty-one per cent. of the interviewees described the reciprocal behaviour which came from the 'understanding' or 'harmony' which has already been discussed in the preceding chapters.

Management, wages, freedom in working

Wages received very little mention, either in a favourable or unfavourable sense. Two of the senior operatives expressed satisfaction with their pay while three of the younger married men resented the wage cut in the six-hour system. A direct question later in the interview confirmed that on the whole there was general satisfaction with wage levels. In this connection the management of the company enjoyed a reputation for fairness in its payment policy which was described as 'very generous' and 'very fair all along'.

There were also references to easy relations with supervisors and to freedom in working which, as appeared afterwards in the new plant, was a factor more appreciated in retrospect. The following extract is taken from an interview held some months after the first transfers and the interviewee himself was waiting to be notified if he was to be re-engaged in the new plant or not. He discussed at some length the initial impressions among his friends who had been transferred:

'I'll tell you one thing, we do enjoy an amount of freedom here in our jobs, call it industrial freedom if you like. There's no regimentation here whatever. You haven't got a staff man on top of you all the time driving you. That is so they tell me at the new factories. I've spoken to quite a few who are there and who have gone from here. According to what they've told me, you have to follow the machinery, you're a robot up there. I've been told by one of them that they go to their jobs there as if they are on parade. It's very boring and very disheartening to them, there's no doubt about that and there's good money and wages there but that isn't everything is it? And in one factory they tell me that every worker there is on the clock, in and out, and you dare not leave your job, someone has to be there to take your place or the process wouldn't work. Now I would like to ask you a question about this clocking in, is it lack of faith by the management in the work people or not? I think there are abuses on both sides'.

Dissatisfactions in the old plant

As might be expected, the physical effort required in the manual operations of rolling and annealing was first in prominence in the negative job features. But in the majority of cases the physical load tended to be mentioned as a statement of fact rather than as a source of acute grievance. For example, there was the second helper who said almost casually 'it's one of the hardest jobs in the country, it's my bread and butter and I find it all right', and the doubler who said 'it's a hard job, you have to be fit. A lot has to do with the hardness of the steel, if you have a soft steel they practically double themselves'.

An exception to this were the men who were generally dissatisfied with their jobs and who introduced a markedly negative emphasis. Complaints of the physical difficulty of the work were accompanied by references to the local men voluntarily leaving the industry in this period of obsolescence, and the recruitment of foreign workers to replace them. An awareness of the current low evaluation in their community and in the Press of the 'donkey work' or 'slavery' of the mills, was another important factor in producing intense dissatisfaction. One furnaceman illustrated these evaluations by narrating how a truck driver watched him working while waiting for his truck to be loaded. Before leaving the driver told him, pointing to the lifting of the steel bars, 'I wouldn't do that job for £50 a week'.

The depreciation of the status of the mill worker in his community and industry was further illustrated by some of the men, using the symbolism of the 'long white apron'. The wearing of the white apron in former decades seemed to represent a golden age in the industry when the hand production workers' social rank was respected. This was especially the case with the older mill workers who expressed dissatisfaction with the diminution of 'pride in work'. They described the loss of prestige of their job and contrasted this with a period when the job 'was an *honour* in my day, you had your aprons and they were kept without blemish all the week'. This symbol of pride in work had disappeared, they said, during the obsolescence of the technology, and latterly the white aprons were no longer worn to and from work.

Although the discontinuation of this custom was probably part of more general social changes, at least they perceived it especially as bearing on the declining state of their industry. Perceptions of the job as low in rank in comparison with the jobs in the new plant, by the evaluations of others or by the employees themselves, were emphasized by the dissatisfied senior operatives as in the example of this rollerman:

'I dislike it entirely. I used to like it, it was the top job, but not now. There are other jobs available in the new works'.

Dissatisfied employees

In this chapter we have been in a sense following the findings of the job satisfaction questions. Now, taking the replies of the dissatisfied employees as a special category, it was possible to discern a convergence of negative comment on inefficiency of plant or equipment: on the obsolescence of the factory in comparison with the technical innovation in the industry. This was accompanied by perceptions of lowered status of jobs which formerly carried the prestige of skills acquired through long experience. The selection of new entrants from the local community had stopped by the 1950's; since then junior production grades were filled by immigrant workers.[3] Thus, the dissatisfaction of the employees in their present jobs was derived from the advance effects of the innovation. To the direct question (How well do you like your job?) a doubler and rollerman made these replies:

'To be candid, I don't like it. It's a hard job and there's no future in it. There's no point in putting your mind on anything better because eventually it will close. I know that practically all the rollermen and doublers would go to the new plant labouring if they had the chance. The rollerman, well, it was the top job. He had authority, now it's changed, times have changed. The old idea was that you'd go on in the mill, work your way on and take promotion and there was prestige attached to the job. Now that's all gone'.

'I don't like it. It belongs to the past. We live in an era now when automation is coming in, this belongs to the past'.

The antecedent effect of the introduction of the new technology produced these dissatisfactions in longer service and senior workers who might have achieved under usual industrial conditions high ratings in job satisfaction: thus the rollermen were among the dissatisfied or least satisfied members of the sample. This finding runs counter to the tendency within a given occupational stratum for upper status levels, higher age or long service accumulation, together or separately to be associated with increased job satisfaction.[4]

The dissatisfaction in the steel workers with senior production jobs was evidently a direct effect of industrial innovation even though its most drastic effects on their skills and employment had not yet occurred. The concepts of stress or anxiety are relevant to these prior effects. Lazarus[5] who has assembled the literature on stress in the individual, concluded that a principal condition for the occurrence of stress required that 'the most important values and goals of the people observed were endangered or disrupted'. Essential to this condition, as Lazarus points out, is the presence of a perceived *threat* which 'implies a state in which the individual anticipates a confrontation with a harmful condition of some sort'. In the case of the steel workers the uncertainty of their future employment presented the threat, against which they had few resources except the reassurance of their existing work group. The period of waiting in the obsolescent factory had evidently produced the anticipation of damaging effects and along with this a diminution in their status as skilled workers.

There were two further possibilities which influenced or modified these reactions: the differences between individuals in their capacity to tolerate uncertainty and in their perception of what the changeover implied. Many workers endured the situation relatively stoically in that they were more able to withstand its uncertainties. Others in the sample who had already been notified that they would be transferred to production or maintenance work in the new factory thought that the new jobs would be more rewarding than disturbing. This extract illustrates the kind of advantage which seemed to them a likely outcome and this sort of anticipatory attitude may also have assisted their subsequent adjustment.

'We've got something to look forward to at the new place if this place is going to close down. There'll be no aprons or tongs, there it will be all push-button'.

These interviews were completed in early 1957. By the beginning of the following year the factory was finally closed and many of its employees transferred to the new plant. It seemed that their degree of adjustment there would be well worth further investigation. It was obvious that the steel workers would be required to adjust into an immensely different job environment and into different forms of social organization with a structure and composition very different to that of the old plant. Later, from the conversations with men who were about to be transferred, the question posed itself: would the dissatisfaction experienced by senior operatives become chronic in the new situation or would other and substitute satisfactions emerge, and if they did what would they be?

Summary

It was suggested that the overall job satisfaction assessment might reflect an individual's acquiescence in his occupancy of a job rather than its positive attractions for him. This will be discussed again when the results after innovation are presented. Altogether, there were three measures which would allow comparison of attitudes before the changeover with those in the new plant: the proportions of satisfied, dissatisfied or undecided employees, the ranking of specific likes or dislikes which were mentioned spontaneously and ratios of favourable to unfavourable opinions expressed by the interviewees (see Chapter 7).

The configuration of likes and dislikes showed that familiarity with the job and work place, environmental conditions, intrinsic job factors, social relations and working hours were each mentioned by more than twenty per cent. of the interviewees. On the negative side, physical work load and inefficiency or obsolescence of plant were most often referred to.

Some of the senior and longer service employees were dissatisfied: this was an antecedent effect of innovation and was also associated with *stress* defined as the presence of a perceived threat. Differences were observed between individuals in their reaction to the uncertainties of their employment future.

REFERENCES

1. Victor H. Vroom, *Work and Motivation,* John Wiley, New York, 1964, p. 50.
2. S. Wyatt and R. Marriott, 'A Study of Attitudes to Factory Work', *Medical Research Council, Special Report Series, No. 292,* H.M.S.O., London, 1956, p. 17. The use of ratios as a measure of job satisfaction is demonstrated in this investigation.
3. *See* J. K. Chadwick-Jones, 'Acceptance and Socialisation of Immigrant Workers in the Steel Industry', *Sociological Review,* **12,** No. 2 (1964).
4. Victor H. Vroom, *Work and Motivation,* John Wiley, New York, 1964, p. 129.
5. Richard S. Lazarus, *Psychological Stress and the Coping Process,* McGraw-Hill, New York, 1966, p. 30.

Chapter 6

The New Technology

There were few similarities to the previous technology in the present rolling, annealing and tin-coating operations. Physical intervention from the operatives in the several stages of the old production methods was replaced by instrument settings to prescribed limits at the start of machine runs of operations. During long periods of the shift the job tasks entailed surface attention and monitoring of machines or control panels.

The contrast to the old machine system was impressive, much of the treatment of the product now took place out of the operative's view and the entire map of production layout was changed. One interviewee told how some retired steel workers, brought on a guided tour of the plant, 'couldn't believe they were in a steel works–they were looking for the furnaces'. The dissimilarity between the old and new plants was very evident in the obvious differences of sheer area and size of the machines, some of which are illustrated in Figure 9.

Continuous-flow production had superseded in this section of the industry, without gradualness, the relatively primitive methods used in hand production: this was the essence of the technological innovation. Those who were transferred between 1956 and 1958 had worked in a plant which, except for a few minor modifications, had not undergone change during the preceding fifty years. They were now in a plant which was unique in Europe for its technical efficiency and production capacity.

We have described how, in the old plant, rolling and annealing processes were carried out on small batches of steel sheets. In the former, powered-machine rolls were controlled by the operatives who lifted, carried and positioned the materials during and between operations. The rate of production depended, therefore, on the human energy input and the quality of the product depended on the correct decisions being taken at each step by the operatives. The so-called hand manufacture was not all that the term seems to imply–neither hand tools nor manual machines, in a true sense, were used. Heavy physical work-load was involved in the loading and unloading of machines, the machine subsequently completing the operations on the

Figure 9. The Cleaning Line of an Electrolytic Tinning Plant.

product. However, the overall frequency, variety or gradation of machine actions were dependent on the operative's decisions.

Limitations on the capacities of operatives to manipulate the steel sheets in and out of the rolls, into the furnaces for re-heating and back again for rolling to a thinner gauge meant that the old factory contained a number of identical machines–each production stage being duplicated and re-duplicated on a row of small machines, each one with its team of operatives and each producing identical batches of the product. Sheets were rolled, cut, trimmed and put through a multi-stage series of processes for surface treatment of the steel–pickling, annealing, tinning–and always with the physical intervention of the operatives. The mobility and activity of groups of operatives in close proximity to each other was thus a main observable characteristic of this type of production.

The new methods consisted of a series of uninterrupted machine operations on one production line. The continuous flow of steel through successive lines of rolls gaining in speed with the thinning of its gauge, was very different from the repetitive passing of bars by hand through a single set of rolls. The nomenclature of these automatic and linked machine operations was broadly the same as before (rolling, annealing, tinning) but there were immense changes in the complexity of techniques and in the general scale of production. The overall output per head in the industry since the introduction of the new methods has been estimated at some ten times that of the pre-change technology.[1] It also meant that the entire tinplate production in Great Britain took place at three plants instead of the scores of small factories which used powered-machine batch methods. The total labour force required for production was considerably lower than before and moreover the ratio of production workers to maintenance and ancillary personnel was very much reduced. So in the new plant maintenance personnel were more numerous than production workers whereas before the changeover production workers comprised over eighty per cent. of the hourly paid employees.

The principal advantage of the continuous rolling and electrolytic coating method was its operation on steel strips moving at high speeds. Figure 10 gives an outline illustration of this production sequence. Steel bars or slabs measuring twenty feet long and five inches thick entered the Hot Strip Mill at five miles an hour and passing through a line of rolls were reduced to strips one thousand feet long and one-tenth of an inch thick, running out at some thirty miles an hour, to be coiled for machine transportation to the next stage. After cooling, the coils, which might weigh up to twenty-five tons, went to the Pickling Line to be cleaned in dilute acid baths, and in the Cold Reduction Mills the strip was again reduced by rolling to a thickness of one-hundredth of an inch, subsequently receiving surface treatment by

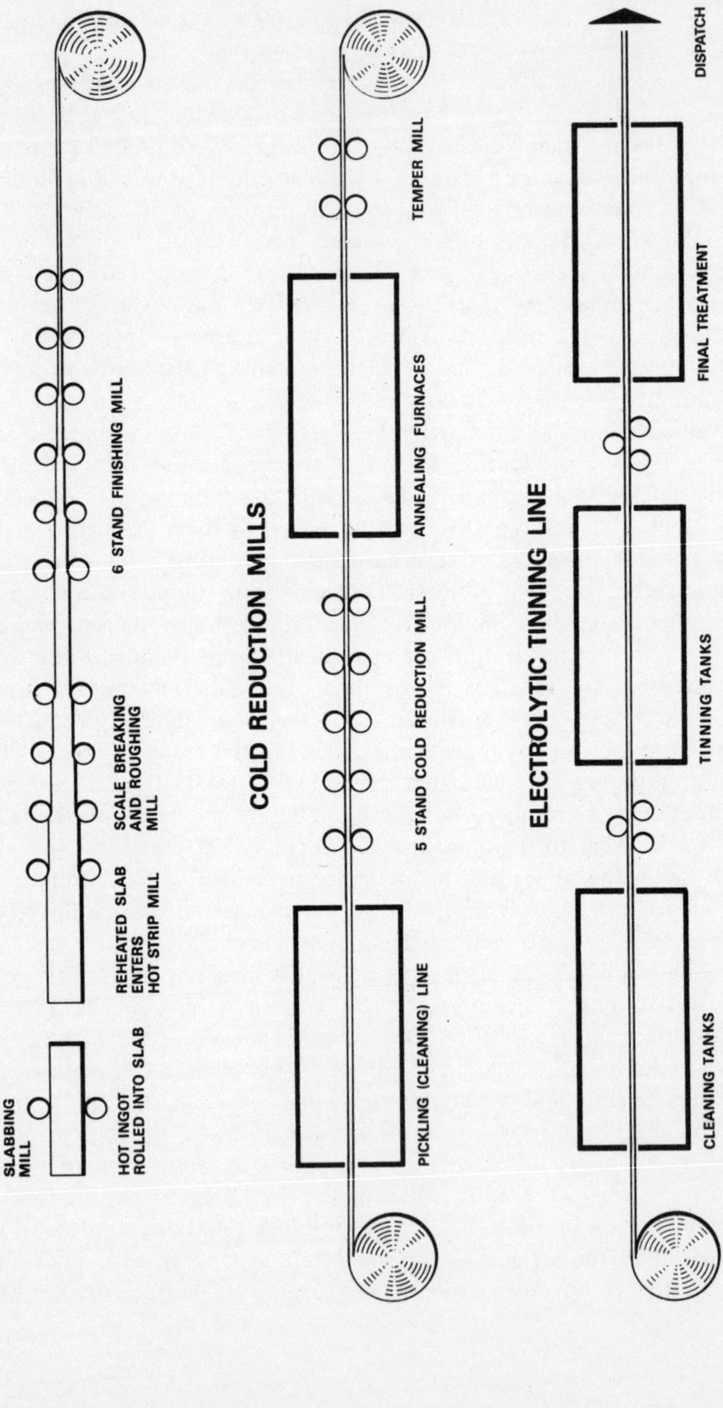

Figure 10. The Production Sequence.

annealing and tempering. On the Electrolytic Tinning Line the strip went through a succession of rolls, cleaning tanks and tinning tanks: here it passed between tin bars in electrolyte solution, a heavy current being used to produce a fine coating on the surface of the steel. Altogether, a series of fourteen tanks were used for cleaning and plating the strip, the latter process disposing of a maximum plating current of 120,000 amps. The final operations consisted of further surface treatment, sorting, cutting and packaging the tinplate sheets for dispatch.

Automation characteristics

Especially characteristic of the continuous-flow line was machine transportation of materials between stages of the process and machine performance of a long series of pre-scheduled operations. As in most cases of automation development, the machine design of the new plant was a combination of components of older technology (for example, power-driven machinery) together with newer elements of programming or pre-scheduling. Applying the taxonomies suggested for the degrees of automation by Bright[2] or by Crossman,[3] this technology would be classified at a more advanced point of automation development than the steel technologies in which human factors were described by Walker[4] and by Scott et al.[5]

This has its human implications. For instance, the strain or 'nervous fatigue' reported by operatives in Walker's study derived from the fact that the equipment 'was not automatic enough'.[6] The contrary was the case here and with a greater degree of machine programming there were fewer demands on the operative. Crossman has pointed out that the increased mental load on steel workers at earlier stages of partial automation was derived from greater overall mechanization and speeded transportation but without an equivalent or sufficient reduction in human information-processing.[7] The changes in tinplate technology, in addition to the replacement of the operative's physical energy, centred on the removal of much of the decision requirements from the employee which, in the batch system, involved the timing of his actions and assessment of the product condition in order to determine the subsequent machine operations. As the degree of automation increases a falling curve might be demonstrated for direct intervention in production by the semi-skilled labour force, and the Electrolytic Tinning Line would be well along this curve.

Let us look more closely at the continuous tinning process. Although the sequence of operations followed automatically from instrument settings at the beginning of the series, techniques for self-correction were not utilized as yet. This was not, therefore, the point of complete machine control but a stage immediately prior to this, where a series of operations was machine programmed and information on performance conditions was relayed to dials for monitoring and intermittent re-setting. Thus a number of the operative's controlling and information activities was replaced by machine action. What were the skills or expertise now required? The roles of the operative in this technology and the type and range of its demands on him, need to be described if we are to include all those features of job content which may relate to the men's satisfactions or dissatisfactions since the changeover.

The production workers

The work routines of most operatives involved monitoring duties to well-defined limits. The senior operatives set the instrument panels to work specifications taken from daily work schedules issued in the production department. They also made settings during the sequence of operations itself on the instruction of the technical or research departments following sample checks on chemical and metallurgical conditions. The operatives were thus required to intervene in operations only to make corrective settings or in order to initiate and terminate the sequence. They were able to stop the line if a fault occurred which could not be corrected by re-setting the controls. Measurements of production conditions were presented in displays which the operatives periodically checked. Their tasks were therefore very largely of a symbolic nature, that is, mediated through dial or display readings and also involved checking the correspondence between readings, although direct inspection of the product for faults was required in a few cases.

These monitoring and symbolic duties drew unfavourable comparisons from those men with experience of the fine judging operations of the mills, such as this one from an annealing operative:

> 'My experience doesn't come into it although I am *called* a process worker . . . (in the old factory) you just had to be alert to the conditions as they varied all the time. The experience was such all the time that you only had to look at something and you could tell whether it was right or wrong and it was generally that you found this, that you had to rectify it yourself. Here that's not the position, all that's taken away from you . . . in the old works you relied on the experience of the work to tell you it was completed. Here, as the clock tells you so the work is completed'.

On the other hand, in the case of 'crash' action where, after machine failures, the line had to be stopped immediately, then some operatives saw the necessity for the anticipation of faults which might cause a stoppage to occur:

> 'It's a job which requires constant attention. You've got to be on the watch, watching all the time and thinking in your mind of the different things that could happen. On the line you can't see the top end from where I am so you are wondering what might happen

to the strip at the top end and how this will affect you down here, the different things that might go wrong'.

The production workers also needed to locate minor points of malfunctioning in order to stop and restart the machine system with as little delay as possible. Where the stoppage was of this nature, the operative himself might get the line going again and in this respect he performed a maintenance task. Thus effective job performance was related to avoidance or reduction of down time on the production line by rapid location of fault origins, and here the duties of production operatives and maintenance staff tended to meet. As the research analysis showed, this part of the operative's tasks had interesting possibilities for the development of individual skills through special training.

Electrolytic Tinning Line

While this outline of operative duties holds for all parts of the continuous process factory, it will be useful to describe the tasks in greater detail. In view of the fact that most interviewees in production jobs worked on the tinning line, an analysis of tasks and skills was made on this section of the process. However, no differences were found between departments in the distribution of opinions on any of the main job topics which will be treated later.

The production team consisted of nine operatives including two whose duties involved inspection of the moving metal strip for surface faults. The wage hierarchy followed the order: operator, shearman, inspector (two), assistant shearman, feeder, piler and electrolyte attendant (two).

An intensive job analysis was carried out on the tasks of operator, shearman and assistant shearman, which was based on a description of job activities of twelve operatives in these three grades carried out during eight hours observation of each: in this way a total period of ninety-six hours of task behaviour was recorded and tabulated. In most instances the observations were included in two basic categories of *signals given* and *signals received* (to or from displays, instrumentation, product material, work mates and other personnel).

Figure 11 shows the work-flow on the tinning line and the locations of the production team of nine men. The operatives were all stationed on one side of the line along a distance of some two hundred feet. They were located in most cases out of talking distance from each other. Even for those in proximity to each other such as the shearman (2) and assistant shearman (3), high noise levels tended to restrict communication. The only exception was at the entry point of the line where conversation could comfortably take place. Technicians came to the production line, for routine quality checks

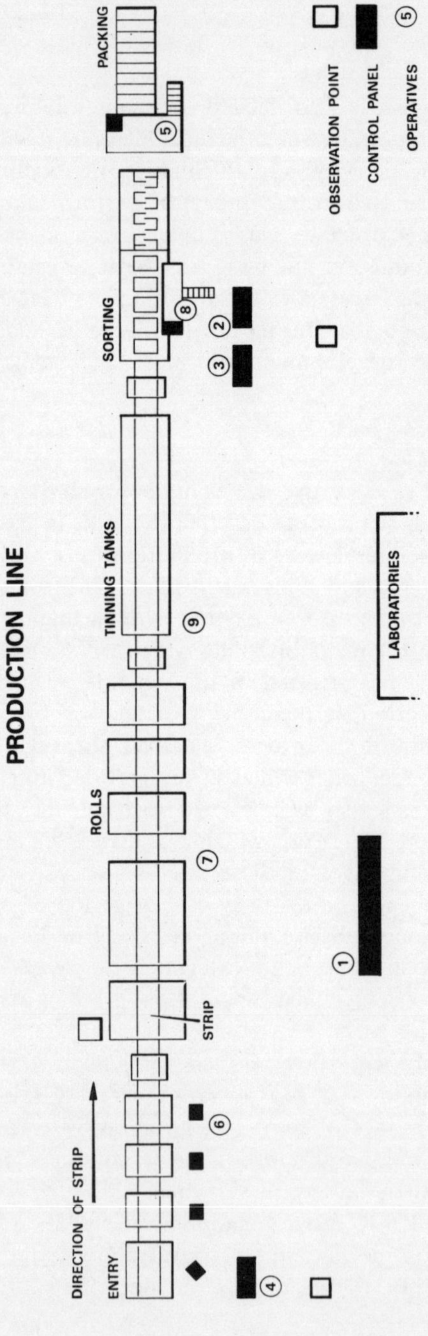

Figure 11. The Electrolytic Tinning Line

and for testing electrolyte samples, from the laboratories at the centre foreground. With innovation in techniques there was an increase in the number of levels of management, of specialist branches and in managerial roles. Maintenance staff were more important both in overall numbers and in function than previously.

The feeder was stationed at position (4), the entry point for coils coming in after tempering and trimming. The senior operative's position was in front of the large display panel at the centre of the line. (6) and (7) were the electrolyte attendants whose duties were to load new anodes into the tanks and to make routine settings. (8) and (9) were inspectors. At position (5) the piler prepared cut sheets for dispatch. The operator, shearman and feeder had a fixed relief break of twenty to forty minutes for each shift, the remainder having shorter periods. During relief periods an operative's job was taken over by the man in the grade below.

Separately from the job descriptions, an observational analysis was also carried out for a period of one hundred hours on interpersonal behaviour of four entire teams, and in Chapter 11 the analysis of communication in the new teams will be stated in detail. One marked change which appeared was that the operative positions did not show the high rates of task-required contacts observed for teams in the old mill. Moreover, the contacts were more likely to occur between relatively distant locations and therefore took the form of brief signals by hand or by loudspeaker. It will be recalled how, before the changeover, both the frequency of contacts and their duration were observed. Task-sharing between operatives lasted from ten to fifteen minutes or more in every hour. Now most contacts were for less than a minute, with only approximately one contact in every fifty extending from three to five minutes in duration.

The job skills will now be analysed, taking the team positions in order of their sequence from entry to exit of the coil. The tasks of the piler and electrolyte attendants were of a comparatively manual kind and need not be described in detail.

The work cycle lasted from the time the coil entered the line to final dispatch, a period of about ten minutes.

The Feeder

The leading end of the entering coil was trimmed square and brought up to the welding station. The feeder entered the coil to the line by operating a hydraulic lever. The end of the preceding coil was then stopped, trimmed and seam-welded onto the new coil. During this operation the coating section of the line was maintained in continuous running because of the length of spare strip stored in loops further along the line. As soon as the weld was completed the entry rolls were speeded up to refill the looping pit. Occasionally the feeder trimmed the incoming coil by hand when there were extremely bad distortions because of damage during machine transportation. He could use control buttons for reducing speed, or for stopping his section of the line should the coil catch while passing into the loops. He signalled the completion of the weld by loudspeaker.

The Operator

The operator's tasks included the greatest allocation of monitoring duties over the machine operations of the whole line. If the line stopped he reported to the supervisor the conditions of breakdown. He alone of the operatives was allowed to restart the line.

Machine setting tasks. The operator set the main control displays to tolerances stated on production schedules received from supervisors. The controls were re-set at the beginning of each new series of coils and on some larger orders these might run on for several hours. The settings related principally to the gauge and the coating thickness. There was also repetitive setting for each coil within the series as it entered the line: increasing speeds and altering coating controls after the weld.

Routine inspection tasks. While the line was running the operator made routine visual checks on the condition of the strip. Observation of the operator's tasks made possible a rough approximation that these visual checks occurred each half hour, with the exception of brief periods of more frequent inspection when technical problems arose.

Detection of faults. There were four common types of fault in the strip produced by the earlier rolling or pickling operations. Action to be taken depended on the severity of the surface marks. The decision whether to add coating thickness or to reject the coil was taken after referring to the supervisor. Another class of faults were those incurred in the tinning line or in the welding operations; correcting them required gross on/off changes in the temperature regulators, alterations of solutions in the tanks, or in voltages used in different sets of rolls. It was sometimes necessary to stop the line when these changes were made. Prompt fault detection reduced the number of reject sheets and it was of obvious importance to limit the duration of stoppages.

Correcting action. With the appearance of faults on the surface of the strip, the operator could make several kinds of machine alteration, for example, to vary coating thickness or roll tensions. Such action might also require that he referred to one of the inspectors on the condition of the strip at some other point of the line. Similarly, it might be necessary to check with the feeder on the condition of coils entering the line. The operator kept records of all stoppages and the factors producing them. In case of major breakdown, he informed the supervisors and maintenance staff and restarted the line on instructions from them. Very occasionally the operator might assist the electrolyte attendant in removing or positioning anodes for changes in coil width between orders.

Location of fault sources. Thus, the operator not only detected faults but, for optimal performance of his job, undertook a search for the possible origins of breakdowns or of surface marking. These might, for instance, be traced to electrical failures, to any of the groups of rolls or to the chemical solutions and temperatures in the series of tanks. Some operators tried a step-by-step search, others attempted random changing of different controls. Since they had no special training, operators relied on a store of practical knowledge and accumulated experience of the line operations. One of them described how he had been a shearman with an operator 'who had become a thinker. After a stoppage you think, it could be either A, B, C, or D and that is how I approach it'.

Operators had little technical understanding of their jobs. They had adopted work methods which they had found to be effective in the past but might, in fact, be taking the correct action for the wrong reasons. On occasions observed, an operator increased the current in one tank above that of the others in order to remove a surface mark. He could not give a technical explanation of why he did this beyond saying that it was 'the way to get rid of the mark'. The operators' behaviour followed the strategies or regularities which they had developed for locating the origins of faults and differences were observed between operators in the way they made decisions. Had they learned concepts, relating similar classes of fault or had they learned, in rote fashion, the actions that generally 'worked' in the past? Such strategies were necessary, as has been mentioned, to reduce the duration of a stoppage, but they were also used when the line was running in order to solve the problem as soon as a fault appeared and so avoid pressing the stop-button. It was obvious that training in this type of search behaviour would be of value to give the operator an appreciation of ideal strategies, of technical concepts and fault categories to assist his decision making, and of tentative hypotheses about the outcome of corrective action.

The Shearman

The tasks of the shearman and assistant shearman were very similar and can be described together. As the job title implies, the shearman was responsible for cutting the strip at the exit end into sheet sizes required for a given coil order.

Machine-setting tasks. Following the specification of production schedules, shear guides were set to gauge and width specifications. At the beginning of an order samples were taken of the first sheets cut and these were measured and checked by the shearman, with the possibility of modification of the guides. For differences in length cutting speeds were altered between orders.

Clerical tasks. A substantial number of the shearman's tasks were clerical: checking the instructions of production schedules and the notes of the previous shift report or writing memoranda for the packing department. At the start of the shift the report from the previous one was checked with the production schedule. When new schedules were provided by the supervisor, two duplicates were taken by the assistant shearman to the packing warehouse and statistics office; he returned with packing schedules and labels to be attached to packing cases. The shearman copied details of the coil series into the shift report and prepared schedules for the packers. He checked and recorded the number of sheets packaged and the weight of rejects, which were packaged separately, recording both in the shift reports. A note was included for the following shift of any unusual conditions of the strip.

Inspection tasks. The shearman inspected the strip and took samples of sheet, passing on details of faults to inspectors. He also inspected the boxes for the general appearance of the piled sheets. If an unsatisfactory weld was made, this was signalled by loudspeaker from the feeder and the shearman could reject the length of strip affected. Routine monitoring was carried out for each of the weld sections of the strip. The shearman informed the operator of any faults appearing at his section of the line. As in the case of the inspectors and the feeder this was usually done by loudspeaker.

Two members of the team, the top-side and the under-side inspectors, observed the strip for most of the shift. Their inspection tasks were identical to those described for the operator and shearman in detection of several main kinds of fault.

How difficult were the new jobs?

Both operatives and supervisors were agreed that there was little difficulty in line jobs and that they needed a relatively short training period. Operatives tended to make the comparison with the hand mills; they held that the new jobs were less difficult than the 'art' and techniques of the former rolling, pickling, annealing or tinning operations. Managers and supervisors were asked to rate the jobs on a five-point scale in training, skill or proficiency requirements. The latter were held to be of general rather than specific kinds. Even for the senior line positions supervisory estimates of skill ranged between 'hardly any' to 'a moderate amount'. Further questioning on training showed agreement that out of several alternative periods of weeks, months or longer, around one month would be the maximum learning time required. It was thought that only one or two of the senior positions in the plant might need more than this. The following excerpt from an interview with a production operative illustrated the common view among operatives as well as supervisors that little training was needed for the actual job operations of normal running:

'Here you could go up and within a few hours you could do any of the jobs there. That's why it's so easy to set up these plants in Japan or in any of the other countries, they don't have to train. they don't need trained labour. Look at my job. I am sitting down in front of rows and rows of buttons, lights and levers. When people come around and see me sitting there with all those in front of me they must think "that job must keep him pretty busy". But there's nothing to it, I only use two or three".

The responsibility of the operative

A five-point scale was also applied on 'responsibility'. Managers and supervisors expressed a similar degree of agreement in their ratings, only on this occasion at the top end of the scale. A 'great deal' of responsibility was thought necessary for efficient performance, and moreover it was required 'most of the time' during the shift. For jobs lower in the team wage-grades where 'hardly any' skill was needed, again responsibility was given as a requirement of importance.

Although supervisors agreed that consequences of error could be extremely serious, they considered the probability of their occurrence to be low. To commit an error was in itself difficult. How real a factor was responsibility where these limiting conditions existed and where human error was not likely to be a cause of major breakdown? Essentially the responsibility referred to was not the kind where errors might lead to damage of costly equipment, rather it was responsibility for reduction of wastage in the product, and was especially related to fault detection and search behaviour. It was true that the production worker took a minor share of this function relative to that of maintenance staff and supervisors, but as was pointed out earlier in the chapter this share might be increased by training methods. Production jobs did not require quick reaction times or visual acuity to any unusual degree.

To recapitulate, the individual's tasks involved his making occasional checks on instrument readings or to detect machine faults (sometimes by auditory signals) which might lead to sub-standard production. While the worker was accountable for the machine setting and corrective actions which affected a large output and cost potential, care and attention in performing simple instrument settings, reporting faults and taking avoiding action to prevent breakdowns, were the full extent of his accountability. Faults were limited to a small number of possible occurrences and there were long periods of waiting when all that was needed was vigilance over a relatively unchanging stimulus field. Thus there were very great and basic differences between his relationship to the process and the continual direct intervention of the operative, in the previous system, as an integral link in actual production operations.

Summary

The main points of difference between the old and new technologies were specified and the automation characteristics of continuous-flow treatment of steel were discussed and considered in the light of taxonomies of machine control. The implications for production workers, their tasks and skills, in the new plant were then described. An intensive analysis was made of the production tasks on the Electrolytic Tinning Line. Most of the jobs consisted of routine monitoring, but the senior production worker was required to undertake search behaviour to locate the source of a fault. Alternative strategies were used by the different individuals who were observed and it was argued that ideal strategies and hypotheses for action could be the subject of training programmes. Other tasks of the team involved inspection, machine-setting and clerical duties. The chapter ended with a discussion of the human requirements of the new jobs. Interviews with supervisors and managers provided, on five-point scales, an assessment of the qualities required in job performance and these results were compared with the description of actual duties. The jobs were rated high in 'responsibility' but low in 'difficulty', and there was no known case of an operative making a serious error leading to damage of equipment. The responsibility of production workers involved vigilance in detection of machine faults to avoid sub-standard production and a general requirement for consistent performance in the context of this kind of responsibility. In the following chapters, when we discuss the men's satisfactions and dissatisfactions in the new plant, we shall return to this point and it will be seen as relevant for the selection criteria which should be used in recruiting personnel for continuous flow production.

REFERENCES

1. W. E. Hoare, E. S. Hedges, B. T. K. Barry, *The Technology of Tinplate,* Edward Arnold, London, 1965.
2. James R. Bright, 'Does Automation Raise Skill Requirements?', *Harvard Business Review,* **36,** No. 4 (1958) .
3. E. R. F. W. Crossman, 'European Experience with the Changing Nature of Jobs Due to Automation', in *The Requirements of Automated Jobs,* Organisation for European Co-operation and Development, North American Joint Conference, December 1964. Final report and supplement, *O.E.C.D.,* Paris, 1965.
4. Charles R. Walker, *Toward the Automatic Factory,* Yale University Press, New Haven, 1957.

5. W. H. Scott, J. A. Banks, A. H. Halsey, T. Lupton, *Technical Change and Industrial Relations,* Liverpool University Press, 1956.
6. Charles R. Walker, *Toward the Automatic Factory,* Yale University Press, New Haven, 1957.
7. E. R. F. W. Crossman, 'European Experience with the Changing Nature of Jobs Due to Automation', in *The Requirements of Automated Jobs,* Organisation for European Co-operation and Development, North American Joint Conference, December, 1964. Final report and supplement, *O.E.C.D.* Paris, 1965.

Chapter 7

After the Changeover: Job Satisfaction in the New Plant

In the study by Charles Walker[1] of the effects of introducing automatic methods on the attitudes of a team of operatives in a pipe rolling mill, the main negative reactions in the new situation were confined to an after-period of two years. Because of this likelihood the second interviewing programme was now carried out after a longer interval, when most of the men had been five years or more in the new plant. We could expect the disappearance of difficulties particular to the early stages of the transfers or of dissatisfactions which were especially initial effects, and the situation would have stabilized to give a full comparison of settled work and social behaviour.

Management and the unions agreed to support further research, this time of all the employees who had been transferred from the old works to one of the new factories, and the number (51) was large enough to give a worthwhile basis for comparison with the previous sample.

The procedure used was similar to that of the first occasion; the men were interviewed in an office inside the factory over a period of six weeks. Interviews usually took an hour or more, and a few of up to three hours' duration were carried over from one day to another because of the quantity of the free-answer contribution. The interview procedure was as in the first sample: the men were asked about the changeover and their own experience of it. Assurances were given, as on the previous occasion, of conditions of anonymity, that the material would be used mainly for the purposes of publication from the University; recommendations to the company would be made in terms sufficiently general to avoid identification of individuals.

Initial questions focussed attention on the job itself but later in the interview a number of specific comparisons were invited on aspects of the job and the work environment.

The men had been engaged at the new plant and assigned their jobs without formal training or induction. Forty-one per cent. were in jobs on the production line although none occupied the one or two top positions in the operating teams. Twenty-one per cent. were with the maintenance staff, twenty per cent. were in ancillary jobs and eighteen per cent. were labourers.

Table 4. Interviewees' jobs in the new plant (*N* = 51)

Production		*Maintenance*		*Ancillary*		*Labouring*	
No.	*Per cent.*	*No.*	*Per cent.*	*No.*	*Per cent.*	*No.*	*Per cent.*
21	41	11	21	10	20	9	18

The majority were production workers before the transfers, two were skilled craftsmen (electricians) and transferred into similar jobs in the new plant. Again, most of the men were married with families and had spent their working lives in the industry. Almost half had been employed in the new plant for six years or more and well over two-thirds of the group had been there for more than five years. Table 5 shows the distribution of length of service since reorganization; the average period was five years and three months. Some of the men had undergone a period of unemployment, up to two years in one case, after the closure of the old plant. Of the five with three years' service or less, three had worked in other pre-automated steel plants and two had tried employment outside the industry before coming to the new factory.

Table 5. Length of service at the new plant (*N* = 51)

	No.	*Per cent.*
Six years or more	23	45
Five years or more	16	31
Four years or more	7	14
Three years or more	3	6
One-two years	1	2
Less than one year	1	2
	51	100

Attitudes in the new factory

We shall now examine the configuration of favourable or unfavourable attitudes in the new plant as these were expressed spontaneously. The interview began with standard general questions which allowed the interviewees to raise the topics or features of the job and its environment which appeared most important to them. The rank order of topics by frequency of mention is shown in Table 6.

Table 6. Job satisfaction in the new plant (*N* = 51)

Likes	*No.*	*Per cent.*	*Dislikes*	*No.*	*Per cent.*
Less physical effort	20	39	Shifts, hours	18	35
Interest in the process	8	16	Boredom, monotony	11	22
Wages improvement	7	14	Social relations	8	16
New skills	4	8	Status drop	5	10
Intrinsic job factors	3	6	Less physical fitness	5	10
Welfare amenities	3	6	Less freedom in work	4	8
More leisure	3	6	Limited promotion opportunities	2	4
Management fairness	2	4	Management distance	2	4
Environmental conditions	2	4	Difficulty in adapting	2	4
Security of employment	1	2	'Tension' or anxiety in the new machine system	1	2
			Environmental conditions	1	2

Less physical effort

The largest concentration of positive opinion drew attention to the undemanding nature of the work: the absence of physical effort in the new jobs was perceived as the greatest gain. According to one boiler attendant: 'There's no manual work attached to the job, it's all automatic, that's one good thing'. Some of those who had been transferred from production to maintenance work saw the reduction of physical load as a principal advantage. A fitter's mate gave his view that 'as far as change is concerned, it is a good thing, there's no doubt about it, there is the better product and the job is much easier, and you are not so physically tired when you get home'.

There was general agreement about the ease of the new jobs, and for those on the production line too the change from heavy manual work appeared to influence their adjustment favourably. The question calling for a comparison in physical effort confirmed that the new jobs were very much lighter, without exception, and in answer to a further question on what was the *main difference* between the present and previous employment sixty per cent. again drew attention to the absence of heavy work. As one of the most predictable and self-evident aspects of the changeover, this finding reflects the objective change in the task requirements: the elimination of lifting and direct manipulation of product materials by operatives. In the report of the European Productivity Agency on 'Steel workers and Technical Progress' this was shown as a major trend by the results of research in six countries on technological change in the industry, the authors concluding that 'most marked of all is a general recognition of a decrease in physical strain'.[2]

Occasionally, the absence of physical effort was seen by the men as a disadvantage because of its coexistent effects of inactivity or lack of variety in the job. This was the point at which the removal of heavy manual work extended beyond its sheer advantages and now the response of boredom or even complaints of poor general health began to be made: twenty-two per cent. of the interviewees complained of boredom and ten per cent. of a decrease in their physical well-being. We will discuss these negative factors again in the following chapter. They were as much the special problems of the automated plant as heat stress or physical load were problems before the automatic equipment was introduced.

Interest in the production methods

The second most frequent favourable characteristic was the interest which interviewees expressed in their job and the production system as a whole. Even assistants in the factory store-room found an interest in the varied machine components which they scheduled to different departments. Eight of those interviewed gave interest in various features of the technology as their main reason for liking the present job. Favourable attitudes towards the physical ease of work did not reveal anything about the job in its own right so much as a relief of the negative features of the old job. Now, on the contrary, the elegance of the techniques of production supplied a source of attraction which was new to the automatic process.

One former shearman pointed out that his present job as machine attendant was 'more interesting, shearing was repetitious work although you were responsible for the plates. This is far more interesting, the various systems that have to be kept under control'. A former first helper, still a production worker in the new plant, thought it 'a little more interesting than the other. In the old works it was the same five-part system day in day out. Here you have the different gauges, different tensions, widths and all that'. The possibility of failure in the machine system was also mentioned: 'if the automatics fail, and you have to watch for this, you have to operate the dials and switches. There is always something to watch'. One line feeder, who was previously a rollerman, tried to introduce an additional challenge into his job by replacing with his own judgement some of the minor machine operations: 'there's plenty of interest in it especially when you are changing sizes. You think, how many turns you will give it to see if you can do it without measuring and it is very interesting to see if you can first time'.

General interest in the complex machines was associated also with a broader view of the process which was possible in this technology as opposed to the narrow focus which has been reported for workers engaged in the fragmented tasks of assembly-line technologies.[3] This result suggests possibilities for job satisfaction which might be encouraged by the design of retraining schemes to include instruction on the machine system with explanation of its techniques.

New skills and responsibility

The acquisition of new skills in jobs on machine maintenance had a strong attraction for former production workers who had become fitters' mates, as these extracts illustrate:

> 'I like it, I must say, because it is allied to my hobbies. To some people electricity is a sort of mystery, if you are interested in a job and you take it as a hobby, you know what is happening'.
> 'I try to be mechanical. I have a motor bike and I try to do all my own repairs. You are bound to come into it then, into the mechanical side, there is always something to learn'.

Besides those mentioning technical interest or the new skills there were three operatives, two crane-drivers and a boiler-room attendant, who found satisfaction in the degree of responsibility for the operations (intrinsic job factors) which their jobs carried. For example, the attendant explained that it was 'something to do with my temperament, it is a job I'm wholly responsible for, I work on my own'. Responsibility appeared to be more valued by workers whose jobs required them to work completely on their own and production workers on the line in fact did not mention it.

The only other component of the immediate job environment or of the actual job requirements discussed by the men was the negative one of lack of variety in the operations. Thus, the proportion referring to positive *job content* factors such as responsibility was perhaps lower than might have been initially expected although, on the other hand, the generalized interest in the machines was both a new and a promising development.

Wages improvement

The third main satisfier was an improvement in wage levels, which fourteen per cent. of the men emphasized. When a direct comparison on wages was asked for, later in the interview, seventy-four per cent. thought wages were better.

Table 7. Are your wages better or worse or just about the same as before? (N = 51)

Better		*Worse*		*About the same*	
No.	*Per cent.*	*No.*	*Per cent.*	*No.*	*Per cent.*
38	74	9	18	4	8

It was clear that nominal wage levels were higher for most men than before the changeover, but the men's perception of the improvement also involved their ideas of current living costs and their knowledge of wages earned by other groups in the industry or outside it. Relatively more hours per week were worked at the new plant, and this led twenty per cent. of those who said that wages were on the whole better to qualify their opinion in view of the necessity for overtime or weekend working. In the case of nine workers who said that they were now worse off than before, these considerations were again mentioned and, more significantly, all but one of them had undergone transfer downwards from production grades into labouring jobs.

On the whole, the interviews reflected objective benefits of a clear and substantial kind. Light work and higher wages, together with the general interest of the continuous line, were demonstrated to be salient satisfiers. Occasionally, responsibility for job operations or, in the maintenance staff, opportunities for acquiring new knowledge were mentioned. Although there were several other topics, the number referring to each was comparatively small or the topics self-explanatory. A few mentioned the company social club and the increase of time available during the week for their personal hobbies because of the new shift arrangements which allowed free days during the week instead of at the weekend: this shift change will be discussed in the next chapter. Improved working conditions received only slight mention as satisfiers, although when the comparison question was presented to forty of those interviewed, eighty-three per cent. said they were better off than before.

Table 8. Would you say the working conditions here were better or not so good as before? (*N* = 40)

Better		*Worse*		*No difference*	
No.	*Per cent.*	*No.*	*Per cent.*	*No.*	*Per cent.*
33	83	4	9	3	8

Most of the men described the cleaner general conditions of working: 'I could wear my best suit here'. Those who thought that conditions were worse complained of poor ventilation; and a further five men who on the whole thought conditions were better, also drew attention to this grievance.

Overall satisfaction: the comparison

Tabulation of self-assessments of the degree of liking of the job shows that compared with the previous sample there was now a higher proportion who were satisfied.

Table 9. How well do you like your job?

Interviewees classified as:	*Satisfied*		*Undecided*		*Dissatisfied*	
	No.	*Per cent.*	*No.*	*Per cent.*	*No.*	*Per cent.*
Before the transfers (*N* = 47)	33	70	5	11	9	19
In the new plant (*N* = 51)	40	78	1	2	10	20

Satisfiers which appeared only in the post-change situation—the ease of jobs and higher wage levels—may have assisted in producing a favourable overall attitude towards these jobs. But there were several other factors which were likely to have contributed. Anticipation of redundancy or uncertainty about their future tended to lower job satisfaction levels before the transfers, and in these conditions some men had become disillusioned about jobs with which they had earlier in their lives been satisfied. The percentage of employees expressing a favourable opinion may therefore have been increased merely by the withdrawal of threat forces of this kind.

There was clearly a conscious effort on the part of the individual to make an adjustment to his present occupation, involving his acceptance of new rewards; foregoing or surrendering what were habitual former satisfiers, i.e. factors in the previous jobs which he had perceived as important. The majority of the workers to whom I talked had made efforts to find advantages in the current situation, although not necessarily easily or without cost to themselves. As one labourer explained: 'adaptability, I think, is important. We were told that when we applied for the job. It was what they were looking for, and there's no doubt that you must resign yourself and adapt'.

Acquiescence or Satisfaction?

Another interviewee, complaining that there were few promotional opportunities in the new job, illustrated how acquiescence rather than satisfaction might be a more suitable description of the adjustment he had attained.

'At the time I came here I was thirty-nine which was not too old for this type of work. I was used to the idea of promotion by ability and working yourself on, but that hasn't applied here. I have settled as *much* as you *can* settle in this job'.

For some employees, their *job satisfaction* was therefore a reflection more of their general acceptance of the changeover than of positive attitudes towards rewards of the job.

There were persuasive influences on the individual towards attaching value to his present situation, which derived partly from the policies administered through the personnel department of the firm—as has just been mentioned the transferees were informed of the importance of 'adaptability'. Moreover, the lack of alternative opportunities for employment was evidently an important constraint towards adjusting expectations to the present jobs. This absence of opportunities was real enough for six of the interviewees to have experienced varying periods of unemployment after the old factory closed before their re-engagement.

Particularly in the early months there were problems of learning how to cope with the unfamiliar and of acquiring new work habits; it can also be assumed that during this period the individual modified his expectations of what the job should provide. The resolution of earlier difficulties was illustrated in these two extracts where the interviewees explained how they had overcome dissatisfactions experienced immediately after arrival:

'You can't pick and choose and you get adapted, but to be frank and honest with you, when I started here I was very discontented'.
'I prefer this now, after adapting yourself to it. You have got to adapt to conditions'.

The majority of the men interviewed had assimilated even the negative aspects of the changeover to the extent that they said they liked their job, but within the category of *satisfied* employees in some cases the balance of satisfactions was barely on the positive side. For example, four of them expressed their feelings in terms of their acceptance of jobs which seemed to have few positive attractions.

'It's not a case of liking anything, it is a livelihood now. I have got to accept what is here for me . . . as a matter of environment, it's what you get accustomed to'.
'We have no option. I mean to say, all the works closed down so we are fortunate to come here . . . you can't say you dislike it, because you have no option'.

This small proportion of minimally satisfied workers does present a limitation on the degree of improvement in overall job satisfaction represented by the self-assessments in Table 9. It is a limitation which is supported by the results of calculating the like/dislike ratios for the interviewees before and after the transfers: the larger number of dislikes expressed in the post-change interviews produced an interesting difference. The ratio for the pre-changeover sample was 0·57 (likes to likes plus dislikes), in the second interviews this dropped to 0·47. Additional evidence was presented by the answers to a direct question on preferences between present and previous jobs.

Table 10. If you were given the choice, which job would you prefer, the present or the previous one? ($N = 51$)

Prefers present job		*Prefers previous job*		*No difference*	
No.	*Per cent*	*No.*	*Per cent*	*No.*	*Per cent.*
28	55	19	37	4	8

While fifty-five per cent. preferred their present occupation and for a further eight per cent. there was no difference, on the other hand thirty-seven per cent. would still prefer their previous job, and this proportion included ten of those who said they were satisfied with their present jobs.

It can be concluded that while the effect of the changeover had been to increase rather than decrease the proportion of those who expressed a general satisfaction with their jobs, the pressures to acquiesce in the transfers tended to mask a reduction of rewards in the new situation. The problem for the transferees did not devolve from any overt rejection of the changeover, but within the context of their general acceptance of it certain problems existed. They might have been problems of the way the employees were introduced into their new jobs or they could involve the objective characteristics of these or of the environment. We can now examine these possibilities.

Conclusion and summary

The men were interviewed again after they had been in the new plant around five years. The free-answer method was used to collect information on likes or dislikes in the present job and the frequency of topics was tabulated. In this chapter the positive features mentioned by the men were described and illustrated from the interviews. The jobs involved very little physical effort and this topic drew most of the favourable opinion. It was seen as a relief from the negative pressures of the previous work–on the other hand, the interest in the complex machinery (second in rank order) and in learning technical skills (fourth) were essentially satisfiers which were special to the new technology. Responsibility was not mentioned by those on the production line although, in a few cases, it was given value by employees working on their own. That it did not appear of importance to the former may have derived from the conditions of the line operations discussed in Chapter 6 where the probability of error in operative routines was found to be low. An association was even more likely with the forms of supervision to be described later, which restricted the autonomy of production crews.

Comparing the overall proportions of satisfied and dissatisfied employees (Table 9) it was revealed that the number of satisfied workers was greater in the new plant; this result might have been influenced by the removal of threat forces and the uncertainties preceding the transfers. In addition, the inducements on employees to accept their new jobs were discussed: it was suggested that *acquiescence* was more appropriate than satisfaction as expressing the feeling of some employees, and in this respect it was of relevance that ten of the *satisfied* category continued to show a preference for their previous employment.

It was clear that the inevitability of the changeover was not questioned by the steel workers who in most cases had shown a readiness to work in the new production system–if they were given the opportunity. Some of the employees were dissatisfied not because they were transferred to the new plant but because of their demotion from skilled production jobs to labouring or ancillary grades.

The finding that the proportion of employees expressing overall satisfaction had become higher after the move from the obsolescent plant to continuous-flow production does not in itself help us to understand the processes of adjustment in the individual or whether he has been optimally matched to his job or the job to him. We will therefore examine in the next chapter the detailed breakdown of dissatisfactions in order to throw light on what factors were involved in adjustment. In Chapter 9 it will be shown that for the individual the latter consists of complex successive stages and these will be explored in the ensuing discussion.

REFERENCES

1. Charles R. Walker, 'Life in the Automatic Factory', *Harvard Business Review,* **36,** No. 1 (1958), p. 112.
2. 'Steel workers and Technical Progress'. A Comparative Report on Six National Studies, *European Productivity Agency of the Organisation for European Economic Co-operation,* Project No. 164, Paris, June, 1959. p. 29.
3. Charles R. Walker and Robert H. Guest, *The Man on the Assembly Line,* Harvard University Press, Cambridge, Mass., 1952.

Chapter 8

The Enduring Structure of Dissatisfactions

Initial effects of the transfers

Presumably, the individual, after his transfer to the automated process, has passed through a period of learning of unfamiliar tasks. As well as these basic requirements of the new technical operations there are, simultaneously, social routines to be learned towards other employees and those in superordinate positions. It is going to be assumed, until this assumption is discussed more fully in the next chapter, that after some initial learning has taken place, particularly of technical procedures, then the employee becomes 'settled' enough in the new situation to enable him to make an appraisal of its advantages or disadvantages. It will also be considered how, with the learning of new routines and the solution of early difficulties of this kind, the transferee's attitudes might have become more favourable towards his new job. If there is such a process—even by the mechanisms of inurement to negative pressures as postulated by Baldamus[1]—then it would have ample time to take place among the fifty-one men after the period of years had elapsed.

The initial period of employment was characterized by difficulties deriving from the circumstances where the transferee's knowledge of the new automated system was very limited and his control over the consequences of his actions was uncertain. Misgivings about the technical operations, experienced soon after arrival were recalled by a production worker: 'working buttons and little levers you didn't know what was going to happen next when you came here first'. A maintenance fitter explained how 'the work is very much easier but it took us a long time before we could get used to it. Not exactly the work but the anxiety, the machines, and being afraid that you were not going to do so well'. Without formal training schemes, this apprehension in face of the technical system appeared to fade during the first year of re-engagement, and disappeared with the development of habitual responses, as the first interviewee now emphasized, 'the buttons and levers do all the work for you'.

The operatives seemed to have attempted to reduce any clash between their expectations carried over from the old jobs and the incentives and disincentives which the new jobs provided. We have seen how in the majority they expressed satisfaction with their present jobs and to this extent they had adjusted. But it did not follow that what were at first perceived as negative incentives subsequently became invested with positive meaning. The men continued to experience a number of negative features in their environment, and the cognitive process of reorganizing their expectations to the available rewards had been only partial insofar as there were a number of dissonant areas in the new behaviour requirements. The dissatisfactions to which the interviewees drew attention followed a pattern which was tied to the perception of satisfiers as they had existed in the previous jobs. This appeared in the form of comparisons which expressed *more* or *less* of certain incentives pertaining to the previous situation. Let us examine this pattern.

The reversal of satisfiers

The dislikes in Table 6 showed an outcome which can be termed a reversal of satisfiers, i.e. some of those aspects which had been prized features of the job now moved to the head of the dissatisfactions. The most frequent grievance was the change to the alternate-pair shift system. Only thirty-three of the total number of interviewees were actually employed on rotating shifts, so that the proportion of those with experience of the shift arrangements *and* who were dissatisfied was even greater than Table 6 would indicate. Responses to the question asked later in the interview on *preferences between the present and previous jobs* corroborated this result. Longer weekly hours and continuous shifts over the weekend were given as their main reasons by those who said they would still prefer the previous job.

Shift work and hours

The new continuous shift schedules which cut across the weekend were organized in a rapid change series of alternating pairs, with two shifts each on mornings (6 a.m. to 2 p.m.), afternoons (2 p.m. to 10 p.m.) and nights (10 p.m. to 6 a.m.), and an extra third shift following in turn after each pair (2-2-3, 2-3-2, 3-2-2). The different teams following this rota had a seven-day series of alternating shifts followed by two rest days and once a month by three days, the latter coinciding with a weekend; employees therefore worked either on mornings, afternoons or nights over three weekends out of four. With work teams on a series of overlapping weekly schedules of this type, the arrangement enabled the plant to work without interruption and ensured the maximum use of capital equipment. During twenty-four hours on a given production line four teams would be allocated to it, one each for the morning, afternoon and night shifts, the remaining team taking its off-duty period. This type of rapid change shift schedule known among the men as the Continental system, took the place of the former arrangement of weekly shift changes, i.e. a sequence of five consecutive shifts on mornings or afternoons or nights and with the weekend free.

The variation in the hours of shifts before the transfers is worth description. Either eight-hour or six-hour shifts were worked, depending on market conditions and the amount of plant utilization required. Thus on eight-hour shifts the labour force could operate all the available plant machinery; with six-hour shifts the same number of men operated fewer machines. Implementation of managerial decisions to change to the latter was negotiated with the union branch and involved compensation payments to the men so that the shorter periods were worked at higher hourly rates. Following trade fluctuations, there were long periods when six-hour shifts were maintained. This latter arrangement produced a weekly rotation of four shift periods–6 a.m. to 12 noon (mornings), 12 noon to 6 p.m. (afternoons), 6 p.m. to 12 midnight (evenings) and 12 midnight to 6 a.m. (nights). It was thus possible on this system for the weekend to extend from mid-day Friday until mid-day on Monday.

The complaints against the present system in most cases mentioned interference with social activities. There were a smaller number of complaints against the length of the shift and the overall number of hours worked per week. Interviewees made the comparison with 'job and finish' piece work, an arrangement which had allowed them to shorten the overall length of the work period (see Chapter 3). This latter type of complaint showed some association with the experience of boredom or fatigue which will be discussed later in this chapter.

The following extracts from interviews with two line operatives illustrate something of the strength of unfavourable attitudes towards continuous shifts and the disturbance in community life which followed their introduction:

'In the old works I had my job at my fingertips, I knew what to do. Coming up here, then, you were groping your way to a certain extent but I think what most people don't like is the Saturday afternoon and Sunday working'.

'I find that the system of working, the Continental working, means that you have to give up a lot of things since I've been here. Our social life used to build up on the week-end. All on the weekend. Here you seem to be more as individuals, that you have your days off in alternate weeks. The general system around means that your mates working in other places have the weekend idle'.

Those interviewed emphasized the loss of comradeship from leisure-group meetings, such as the regular 'Saturday night out'. Quite simply it meant a curtailment of friendship relations: 'I've had to give up a lot of the friends that I used to have. We used to go to football and things like that, and I can't always meet them now, friends and members of the family'. Two interviewees only mentioned the inconvenience of altered meal times which 'could be upsetting to the constitution'. One man also referred to his church attendance: 'that was my social activity, church work, and I miss that very much'.

There is no doubt at all that the employees who were transferred from weekly shifts with rest periods at the weekend to continuous shifts with rest days occurring on one weekend in four, disliked the system for its effects on their activities and interests outside of work;[2] it was not so much the alteration in domestic routines which created the difficulty, but the loss of the weekend for social activities. This long term non-acceptance of the shift schedules takes the concept of adjustment on to a much more prolonged time scale from that suggested by the earlier studies on shift effects[3], and forces our attention towards a different level of analysis than that of physical adaptation or domestic and mealtime routine taken only as a matter for individual readjustment. In other words, what emerges is non-acceptance by the group rather than by the individual, and this implies inter-relatedness of changes inside the industrial firm to the general social environment of the enterprise. The delay in adjusting to an in-plant change in shift arrangements derived from an outside source: from the necessity for incorporating the change into the general social behaviour outside the factory. Presumably acceptance of shift work would take place only over a period during which a new pattern of leisure activities could develop in tune with the new shift rotas. In the case of the steel workers it would seem that behaviour expectations in the community where they lived, concerning the leisure group activities of the weekend, prevented or delayed their acceptance of continuous shifts. These perceived social deprivations persisted as long or much longer in some instances as dislikes encountered within the factory.

Continuous shifts brought about a pattern of leisure which was centred more on the home and on the nuclear family unit. For example, the men's wives did not share the views of their husbands about shifts: almost all of the thirty wives who were interviewed found that the transfers produced a change for the better in their own lives and most frequently because the husband now 'has more time at home'.

Whatever the eventual implications of this development or its positive aspects, at the moment it was clear that the steel workers had suffered from shift arrangements by the break created with the wider leisure group. There had also been some influence on social relations *inside* the factory from the same direction because it was now more difficult for men from different shifts to meet during the weekend as they did in the pre-automated technology, and a build-up of credit in positive social relations for the work group, through the additional ties of friendships formed outside of work, was unlikely. Shift work was the reason most frequently given for non-attendance at union meetings which previously were always held at the weekend. Out of forty-five men who were asked the question on whether they went to union meetings more or less often than before thirty-one of them said they went less or not at all, and of this number eighteen gave shift arrangements as a major discouragement (See also Table 24, Chapter 11).

Dissatisfaction with shift work was not a matter of individual preference or of the physiological effects of the new arrangements of working hours. A variety of customary social memberships were interrupted and frequent regular attendance in leisure associations was prevented. It must also be recognized that the individual's adjustment to shift working was dependent on the extent to which the community life around the factory focussed on activities of the weekend.

Boredom and tedium

Boredom or lack of interest were major dissatisfiers in the automated plant. While the interviewees expressed a range of subjective states, it was possible to discern some common trends particularly towards the referents of the responses. Most of those who drew attention to this problem were production workers, who stated the problem in somewhat convincing terms, as is exemplified by these answers to the interviewer's question: *How well do you like your job?:*

'Very boring. It's a good job, in comparison to the work in the other factory, there's no comparison at all. Here it's light work, actually. Another aspect we see of it especially in our department is that it's a dead end job. It will never be any different to what it is at the moment. The tendency of this job is to be protracted boredom, as regards the work. There's nothing in it that you can say "I can take a real interest in the work because I am part and parcel of it"—which was our system in the old works'.

'Financially I am much better off. I've lost my weekend, but I've gained just by the money that's all. Sometimes it can be very boring. It's dreary'.

'The job is so simple that it can be termed "just boring". It requires good eyesight and quick thinking, quick reaction rather than quick thinking, but that becomes automatic after a few years. You find you can do it without really concentrating. It's so simple that you can't boast or brag about it'.

These responses clearly emphasized the tedium of watching the line or else the low difficulty level of tasks: these possibilities could together or separately produce the boredom effect. The problem seemed to be basically that of sustaining a continuous surface attention, where the rate of signal input to the operative was not sufficient to maintain his interest. The salience of boredom in the configuration of dislikes pointed to an important problem of the automated technology and a disincentive which was intrinsic to job content.

A small proportion (six per cent.) of those who expressed this grievance were former production workers who now occupied labouring jobs. These men, for whom the adjustment had been most difficult, were under-employed in a self-evident sense and they complained of the lack of purpose of their jobs. Their dissatisfaction was directed towards the placement policy of the company, because capacities which they had exercised in the senior jobs were not needed. The following remarks make the point that the interviewee felt he was 'tolerated' in the new factory and he considered that his own contribution was very insignificant:

'It's quite an easy job but it's boring. I think we are capable of better jobs. There's nothing interesting in it, there's nothing attractive in brushing the floor, is there? And feeling you are just tolerated and your interest in the product even is nil because you are not part of it'.

Those who had not earlier referred to it as a job dislike were asked a question on whether there was any boredom in the new jobs; the direction of response having been ascertained, the men were then asked for a comparison with the previous job.

Table 11. Is there any boredom in this job? How does it compare with the previous job? ($N = 51$)

More boredom		*No difference*		*Less boredom*	
No.	*Per cent.*	*No.*	*Per cent.*	*No.*	*Per cent.*
24	47	14	27	13	26

Twenty-seven per cent. of the interviewees saw little or no difference in this respect between their present and previous jobs or else were unable to make a comparison because of the qualitative changes in job content. Twenty-six per cent. found in the present job less boredom than tne previous one—these were maintenance workers, crane-drivers or storemen, some of whom found a greater variety in their jobs compared with the routine aspects of the one-hour cycle of operations in the mills.

Forty-seven per cent experienced a greater degree of boredom, most of these men were production operatives or ancillary line workers such as packers. The former complained about the sameness of job tasks, confirming the experience of under-activation which was indicated in the free-answer interview. One operative described the tedium of waiting periods when 'you are just standing with your hand by the stop button'. There were several points of negative comparison with the mills, the most important of which were the restrictions on physical mobility or activity in the line jobs and the absence of the piece-work incentive. The latter was mentioned in the following comparison of the present eight-hour shift with the job-and-finish system. This operative also referred to the variety of job exchanges in the hand mill crew:

'We didn't have time to be bored there, you had your job and you clocked out. Here you have to stay the full eight hours. It's a strain on you, then, you know you have got to be there to do your job and you can't get it over with. In the mills you were having a change of jobs, because of the different "parts"'.

For those who were not production workers, the boredom described involved a more basic feeling of lack of immediate purpose in their jobs and this was especially the case for the former senior operatives now in lower-grade or ancillary jobs:

'The job is more boring, I haven't got any responsibilities you see, so I am not taxed in any way then. As a rollerman I used to carry responsibility, I've lost that now'.

Other insights into the individual's perception of boredom were given in the interviews, where several of the negative effects related to boredom, including fatigue, were mentioned. The similarity of boredom and fatigue in the subjective feelings they produce is well illustrated by this interviewee, both job content and longer term goals are involved in his experience of 'boredom':

'I can say this, generally speaking, there is a large amount of boredom attached to the work. It's repetitive work and there's no change in the nature of the work. You see, in the old works at the start of the shift it was a challenge to you. Here, you accept the thing, there's nothing there to interest you or give you that kind of challenge. It's a kind of drifting in to do a thing mechanically, it's a very difficult feeling to express clearly and precisely. Whilst the arduous nature of the work is very small in comparison to mill work, I believe I am correct in saying that after you go home from work today you feel more tired and lazy, that's the feeling and it can't be attributed to working hard. Another contribution to boredom here. The prospects of promotion are so remote, it was the ambition of everyone in the mills to go on in the grades, to go furnacing or rolling in order to get the money. Here, the position to get promotion is so remote that you think, "Well, I'll never go beyond this job", then you get more discontented and more bored. If the position was better for promotion, it would add interest and keenness to the work'.

Promotional opportunities

Although dissatisfaction with the lack of promotion opportunities was mentioned only in two instances, possibly because it was not an immediate factor in the actual job, it did add to the disincentive effects of boredom. A direct comparison question on promotion gave these results.

Table 12. Do you have the same opportunities for promotion, or less, or more? ($N = 51$)

No. opportunities		*Adequate opportunities*		*Few opportunities*	
No.	*Per cent.*	*No.*	*Per cent.*	*No.*	*Per cent.*
31	61	17	33	3	6

While thirty-three per cent. thought that the present opportunities were adequate, six per cent. saw very few possibilities and sixty-one per cent. were sure that there was a complete absence of any promotional prospect. The aspirations which the men had developed formerly were fulfilled by a promotional ladder of the seven or eight operative grades (see Table 2, Chapter 3). Now, the blocking of these possibilities was not caused by the reduction in number of operative positions, which were again differentiated into a hierarchy by title, duties and wage levels, but by the management policy of introducing younger operatives into top positions when the new factory was put into operation in 1956 and 1957. This prevented advance through the grades such as would usually occur through the retirement of older operatives from production jobs and it made the upper end of the hierarchy practically inaccessible to those in lower grades.

Interpersonal behaviour

Some of the unfavourable opinions reported since the transfers were summarized under the heading of *social relations*. First indications of negative changes appeared in the non-directive part of the interview, following the general question on likes or dislikes experienced in the present job. They arose again in the replies to a number of other questions: those concerning the interviewee's attitudes to his previous job, his preference for the present or previous one, and on the perceived main difference between them. Information from these sources, together with observation of production teams, is presented and discussed in the later chapters. Something of the change was revealed when the men drew attention to the loss of the 'family spirit' and closeness among employees which had existed in the old plant. 'The camaraderie has gone, the men have not come close together like in the old works, it has taken me a long time to weed out who are my friends and who are not'—was a remark typical of this class of dislikes.

Informal reciprocal roles between members of work groups did not exist at the previous levels of intensity, and the inference can be made of a decrease in supportive behaviour from other workers with whom these interviewees were in contact. Influences which would work against the continuation of the closeness which was characteristic of the old plant were the increased size of work teams[4] and the locational separation of their members. These are no more than suggestions here which will be followed through more fully later when I shall discuss the nature of the development of interpersonal behaviour within the plant and how far it was considered rewarding by the individual.

Status depreciation

Here, we take up the threads of the analysis in Chapter 5, where the impact of change was observed to have already modified the self-perceptions of senior workers, producing dissatisfactions particularly for those who had held the most skilled and responsible jobs. This pattern was continued and acerbated in the case of those whose transfer to the new process had involved an extreme status drop, from head of a production team to labourer. They were a small minority (twelve per cent.) of the men interviewed, but they provide an elucidation of what can be inferred in less visible or drastic forms for many others. Adjustment could here be seen making its heaviest demands on the individual. These men tended to 'talk through' the different stages and difficulties of their acceptance of the transfers, and it was possible to follow the gradual modification in their existing values to the point of their partial commitment to the new situation.

Perhaps the best way of interpreting this process is to conceptualize it as a reduction of imbalance or dissonance between the individual's expectations and what the job does provide. Festinger's well-known work on cognitive dissonance has suggested that this process 'centres around the idea that if a person knows certain things that are not psychologically consistent with one another, he will, in a variety of ways, try to make them more consistent'. Dissonance is defined in terms of a person's expectations, 'in the course of our lives we have all accumulated a large number of expectations about what things go together and what things do not. When such an expectation is not fulfilled, dissonance occurs'.[5] Using this interpretation, the problem for the former rollermen and other senior grades was in its essence how to reduce the discordance or dissonance between their expectations or self-perceptions and the actual requirements of the jobs they now held. Since nothing could be done to alter the external situation, some degree of perceptual reorganization had to take place—according to this line of theory.

Something of the process of adjusting or modifying expectations can be seen in this extract from an interview with a former rollerman, aged fifty at the time of the transfers, now a labourer:

'As I feel now I am like an old race-horse put out to graze after a great and glorious past. As a rollerman I was something but here . . . I had come on through experience and in all modesty I knew my work and really had an interest in it. And if you had an interest in it, it meant you were producing "primes" rather than waste and you had that pride that you were able to do it, and again you had to carry the can if there was any waste or breakages. You were the rollerman. And what I may say, the dignity of it, there was no regimentation about it, you were there, a compact happy crew, and as long as the rollerman did his work, you weren't bothered by anyone. I feel as if I have come to the crossroads, as a steel worker. There's a dignity in labour and I put all I could to have the "primes" and now it's finished. Sweeping the floor doesn't have much to do with producing steel. I've had my lot. The rollermen in the new mill get more than £35 per week, I could be envious but I just put it out of my mind. I've got to accept it'.

In cases such as this one, innovation had required the individual to make adjustment of the most severe kind. We have seen examples earlier of individuals who had largely been able to achieve acceptance or acquiescence in their present situation. But for those senior operatives who moved to labouring grades in the new plant, acquiescence had not reached a point where they could state in the interview that they were satisfied in general with the job, even after the prolonged period since the transfers. One rollerman expressed his perception of the change in his job very briefly: 'I can't say I like the job, I feel I am bottom dog here now'.

However, of the thirteen former rollermen interviewed, this degree of status loss had happened to six and the remainder had, in fact, made what they perceived on the whole as a satisfactory move to maintenance or production jobs in the middle of the grade hierarchy.

The extreme status drop occurred because of an age disqualification: those workers in the old plant who were over forty-five at the time of the transfers were not, because of a company ruling, re-engaged as production operatives. Even a cautious inference would suggest that the application of this rule without consideration of individual capacities must have resulted in the misplacement of some individuals. Research has established that effects of ageing on learning or proficiency in skilled performance are a matter of very wide individual variation,[6] and it is very possible that these workers were not 'old' at all for what was wanted in the job.

In Chapter 6 it was shown that the new jobs had required for their performance a general responsibility factor—although not at a high level of difficulty in the actual operations to be performed in normal running of the line. But it was seen that the operative could reduce product wastage by effective search strategies in detecting the origins of faults on the surface of the metal and provision could be made for this in training. The experienced or older worker would, in fact, be quite likely to possess the characteristics necessary in jobs of this nature. It may plausibly be suggested that if a senior worker in the old factory had carried responsibility both for product quality and team performance, it was probable that he would be suited to the new production jobs. In any situation of technological change this grade of worker is liable to suffer most by displacement, and unless there are definite requirements which debar them (on the criteria of research evidence) then the re-employment of such people in equivalent positions may be a preferred solution.

Did management over-estimate the new skills?

The selection procedures which were used for the re-engagement of employees to the new production jobs appeared to have been based on incomplete analysis of job requirements. It may have happened, if management and unions did not appreciate (at the time) the nature of the qualities required, that excessively strict or inappropriate personnel selection standards were applied. The following reference by another former rollerman to the selection procedures used by the firm points to the use of methods which may have been of doubtful relevance to the jobs to be filled. For example, a form of oral test was used which required the applicant to answer mental arithmetic questions:

'I would say this, you may think I have a chip on my shoulder, but when we were interviewed in the old works, we were given various tests, mathematical problems. Many people were frightened by them and I think this was a means of keeping the older people out. There are many people from the old works who could do this job I am doing'.

This can be contrasted with the selection policy in a chemical plant described by Helfgott,[7] where the management decided to use tests to select employees for retraining. It was found that 'on the basis of aptitude tests a large number of the employees, particularly long-service ones, failed to qualify . . . '. Later, on examination of the results, management concluded that the failures were accounted for by deficiencies in basic educational background, and took steps to give courses which enabled most to complete the retraining scheme successfully.

Summary

Initial difficulties of interviewees in their new jobs were overcome as learning took place. But certain negative factors continued to be perceived: in this respect, the new shift arrangements; boredom in the attentional requirements of the continuous line; the absence of satisfying social relations, were topics receiving frequent mention. There was a reversal of satisfiers insofar as some of the prized features of the old jobs were now a focus of dissatisfactions.

The transfers brought loss of rank in the grade hierarchy for some interviewees; others complained of little or no promotional opportunity. Use of an age bar on re-engagement to production jobs was discussed in relation to the skills required for their performance. Two points were made. Namely that dissatisfactions were increased among those who were excluded for this reason from production grades, and secondly, that there was no evidence of special skill requirements to justify an age bar. This finding was considered for its relevance to selection procedures in similar situations.

REFERENCES

1. W. Baldamus, *Work and Effort,* Tavistock Publications, London, 1961.
2. For similar effects *see also* W. H. Scott, J. A. Banks, A. H. Halsey, T. Lupton, *Technical Change and Industrial Relations,* Liverpool University Press, 1956.
3. Early studies were concerned largely with the physiological consequences of shifts. A discussion of the results of more recent studies is given in J. K. Chadwick-Jones, 'Shift Working: Physiological Effects and Social Behaviour', *British Journal of Industrial Relations,* **5**, No. 2 (1967).
4. Bernard M. Bass, *Organisational Psychology,* Allyn & Bacon, Boston, 1965, pp. 200-202. Bass suggests that although smaller units are more efficient, management establishes groups considerably larger than the optimum size.
5. Leon Festinger, *A Theory of Cognitive Dissonance,* Harper & Row, New York, 1957.
6. A. T. Welford, *Ageing and Human Skill,* Oxford University Press, 1958. In the course of the presentation and discussion of a very comprehensive collection of data on aspects of ageing and performance, the author emphasizes that 'a rigid age bar would seem unjustified for at least many training courses', p. 268.
7. Roy B. Helfgott, 'Easing the Impact of Technological Change on Employees: A Conspectus of U.S. Experience', *International Labour Review,* **91**, No. 6 (1965).

Chapter 9

Individual Adjustment

It would appear that several processes were involved in the individual's adjustment in the automated technology, which might be specified as being of three kinds or stages. These would consist of, in the first place, the anticipation of the coming change. Secondly, this would be followed after the transfers by learning of new jobs and roles. Thirdly, and parallel or subsequent to the learning process, can be assumed the development of positive or negative attitudes towards factors in the job environment. At some point after the period of learning, a stable evaluation of gains or losses experienced by the individual thus becomes possible and the sequence of adjustment is now completed.

The anticipatory period

The stress of anticipating a number of unknown consequences of approaching innovation in the period preceding it was described in terms of the anxieties which were observed in the workers interviewed before the transfers. Dissatisfactions were associated with the uncertainty inherent in the circumstances of the obsolescent industry and were accompanied also by anticipatory categorizing:[1] i.e. attempts to formulate ideas of what the new jobs would be like, for example, from information passed around by the first employees transferred when the continuous production lines were put into commission. The interviewees' attitudes indicated that they were forming their response to the new jobs in advance, with the requirements of the 'push-button' environment already in mind, as was shown in those workers who expressed the view that 'we've got something to look forward to'. So it was clear that long before the innovatory changes, the adjustment process had already begun, although there was no training or formal preparation for the changeover.

Job learning

During the initial months of engagement in the factory, the 'tension' and uncertainty which were expressed faded with the learning of the job operations. The complexity of the technology, as far as the operatives were affected, was now found by the interviewees to have a face-validity only and to require, for the most part, simple actions. Instruments which appeared at first to involve difficulty in their operation—the 'buttons and little levers'—were now found to 'do the work for you'.

It could be inferred that this period, where the appropriate actions had to be learned in an unfamiliar machine environment, was terminated by the achievement of habitual responses and routines. It provided the initial impact of tensions, which seemed to be similar to the experience of the operatives described in Charles Walker's study, during the early months after the introduction of new machines.[2]

Social learning

Learning of social routines would also have begun in the initial period; this was likely to be more protracted even though it occurred concurrently or merged with job learning. It was, however, particularly important for the individual to achieve social learning or socialization in order to locate stable reference points for his behaviour and in placing himself in relation to the roles expected in the new organization. We might assume that the immediate social group would be of first importance for providing cues or guides, assisting the individual worker to make *categorial* solutions about his own duties and status and that of others.

Evaluation

Simultaneously with the learning of tasks and roles, the individual can also be assumed to have made an appraisal and evaluation of what the job had to offer him generally in its advantages or disadvantages. In the case of the transferee, appraisal was a matter of comparison of better or worse or 'as much' or 'more' of rewards, which were available in the previous technology. Job learning, as we have already mentioned above, resulted in the environment stabilizing to the point where he could perceive the full extent of its present and future rewards: its range of incentives, positive and negative. The general level of satisfaction which was attained also involved an awareness of what was accessible or what it was realistic to hope for in a situation which the individual was under very strong pressures to accept.

The extent to which expectations acquired in the previous job were discarded in the process of the individual forming his new relationships to the work situation, was described in previous chapters. However, there appeared to be a limit to which this could be taken and certain positive incentives of the previous jobs could not be replaced or substituted; similarly certain current negative incentives were not diminished or modified by the necessity to acquiesce in the new environment and organizational structure.

So much of the writing on automation effects has emphasized the necessity of adjustment in the individual, and it is a fairly common and even commonplace statement that attitudes may be inappropriate or that attitude-change should be attempted by the administrators of innovation.

It seems very likely that workers may come to a new technology with inappropriate attitudes and these may hinder adjustment and may later have to be changed. However, some attitudes or expectations are probably related to factors which are more absolute than relative in their nature, deriving from the wider society and referring to the value attached to work and its meaning for the individual. At this point it must be admitted that the necessity for adjustment is now removed from the individual and placed instead on an adjustment in design and of social organization.

Motivational indicators

One of the interesting findings of the comparison questions was the high proportion of employees who claimed to experience greater 'tiredness' even though the work itself was light in terms of physical effort. A direct question received the following distribution of replies.

Table 13. Does your work here make you feel as tired, more tired, or less tired than in your previous job? (N = 51)

More tired		*Less tired*		*Little or no difference*	
No.	*Per cent.*	*No.*	*Per cent.*	*No.*	*Per cent.*
33	66	11	22	7	12

Twelve per cent. of the interviewees saw very little difference, or else, finding the comparison difficult to make, were not able to say more than that the degree of tiredness they experienced was about the same. Twenty-two per cent. said that they were less tired now than previously, drawing attention particularly to the hard physical demands of their old jobs. Some of these differentiated between the physical exhaustion of the mills and a different kind of tiredness which they currently experienced: 'I liked the previous job but now I can see how much strain it was . . . more or less up here it isn't tiredness but laziness'. Storemen and fitters' mates in particular found their work much easier than their former production jobs in the mill crews.

Sixty-six per cent. said that they were more tired now than previously.

Again, the interviewees differentiated between their former physical exhaustion and the present experience of tiredness, as in these two cases:

'I am more tired than I ever felt in the old works. I have no inclination at all, as if your mind seems to be dulled somehow. I don't seem to have any inclination and it isn't physical tiredness, it isn't exhaustion'.
'I don't think I am as energetic and enjoying life as I was . . . you go out from here tired. It's not exactly tiredness . . . not the same tiredness as the mills'.

Interviewees had difficulty in describing the nature of their 'fatigue'. Verbal report has its weakness for the articulation and expression of a state of feeling of this kind. But even so, it is possible to see certain trends revealed by the analysis of these replies.

Responses to the general question on tiredness can be separated into three categories. There were general statements describing subjective feelings of tiredness which were made by the thirty-three 'more tired' interviewees. In addition, twenty-four men made references to the time or occasions *when* this feeling was most strongly experienced. Lastly, twenty-six of the interviewees attempted to explain *why* they now felt more tired.

Descriptions of tiredness

The description of the feeling state was made in terms of what the men called 'laziness' or 'weariness' and they contrasted this with the fatigue of the hand mills. Attempting to describe this sensation, the men made references to circumstances where they observed changes in comparison with their behaviour prior to the transfers.

Seven of the men mentioned sensations of tiredness which they experienced during the work-shift. Four of these also noticed such effects after working hours, and in this they joined the majority of the category who perceived changes largely or wholly in their leisure activities. Compared with the pre-automated situation, they participated less in hobbies or household tasks, went out less in the evenings and generally tended to discontinue previous interests, such as gardening.

The leads provided by this interview material suggest that we are examining intermingled boredom and fatigue effects. The descriptive references to drowsiness or laziness pointed to lack of arousal and absence of stimuli in an environment where the employees became tired as one said, 'through the effort of watching the clock'. The cross-tabulation of responses of fifty of the interviewees who were asked the comparison questions both on boredom and tiredness confirmed that there was a considerable overlapping between these two subjective states in the perceptions of our respondents.

Table 14. The boredom-tiredness relationship (*N* = 50)

	More tired	*Less tired*	*No difference*	*Total*
More bored	19	3	1	23
Less bored	6	4	2	12
No difference	8	3	4	15
Total	33	10	7	50

Explanation of tiredness

The interview material also showed the men searching for explanations of their tiredness. A few complained that they felt tired because of longer hours of work compared with the previous six-hour shifts; occasionally the alternation of shifts in the Continental system was mentioned. Other explanations given were the continual noise of the production line and the fixed work pace as opposed to the variable rates of the old system. There were also hints at changes in the work team which were thought to contribute to the same end result; these will be considered fully in Chapter 11.

'One vast difference is that there was more time to relax, we were more like a family than here in the new works. Everyone would say we were happier. Now we don't work so hard but we feel more tired'.

Another type of explanation which was offered by the interviewees involved the environmental conditions in the new plant. Almost all in the 'more tired' category made explicit references to certain environmental effects which they claimed were produced by the machines. The following excerpt summarizes this body of comment more effectively than any attempt at paraphrasing. Almost incredibly, this type of opinion was widespread at the plant:

'I have got a fancy myself that there is too much electricity here and whether that is taking it out of us I don't know. There must be millions of volts about here and that must be it, I believe, it's taking the oxygen out of the air. You people should know about that. I don't say it's right for a moment, but it's what a lot of us feel'.

Eighteen of those interviewed alluded to these electricity effects: that there were certain atmospheric conditions associated with the high electrical voltages which made them feel tired. How such a belief as the 'electricity effect' could have taken hold is itself a topic which it would have been interesting to have explored further. A thorough training of operatives covering technical aspects at a more comprehensive level than merely that of the immediate job operations would at least have made it unlikely that this particular belief should have occurred. A pre-condition for holding beliefs of this nature was probably a very limited understanding of the technology.

In their work role even production operatives were relatively remote from the machine operations as, for example, on the electrolytic tinning lines. There were cases which were mentioned in Chapter 8 where they pointed out that their participation was minimal, as in the case of the operatives who explained that 'as the clock tells you, so the work is completed'. Ancillary workers who saw themselves as 'outside' the production operations also felt remote from the general goals of the organization:

'You are not part and parcel of the company, you are not adding anything . . . the plant and the process is interesting but you are not part of it'.

The experience of tiredness, then, appeared to be connected with monotony in the job content; with inactivity and with dissociation of the individual from the machine operations in both ancillary and production jobs. We might assume that the motivational consequences of these factors were to decrease the positive attractions and incentive value of the job itself.

It was relevant to the question of the operative's fatigue or activation to examine the answers to a question on more or less 'brain work'. Although a slightly greater proportion said there was more than in the previous jobs, this meant in many cases that they now had to do few physically exacting duties and therefore there was more 'mental' work. The distribution of replies was as follows:

Table 15. Is there more or less brain work in your job now, or is there no difference to the previous one? (N = 51)

More brain work		*Less brain work*		*No difference*	
No.	*Per cent.*	*No.*	*Per cent.*	*No.*	*Per cent.*
22	43	19	37	10	20

Twenty per cent. found either that the jobs required about the same 'brain work' or else stated that a comparison was too difficult. Thirty-seven per cent. thought that there was less brain work in the new jobs; these men were largely former production workers who were now in ancillary and labouring jobs. Forty-three per cent. considered there was more brain work required since their transfers to the automated plant. These interviewees included production workers who emphasized the *relative* importance of 'mental' as against the small 'physical' requirements in their present job. Those attached to maintenance gangs as fitters' mates also emphasized the greater mental component in their tasks. Because the physical load was negligible a greater importance was given to 'mental' work but this did not mean that the latter was onerous.

The retrospective question

The interviewees' accumulated experience of work in the years before the transfers provided a frame of reference for current perceptions—we have had repeated confirmation of this—but there was also an interactive effect inasmuch as the present jobs influenced the men's recall of the advantages and disadvantages of the old. When the questions were asked in the new plant: *How well did you like the previous job?* and *What was it you liked or disliked about it?* the topics mentioned were again mostly the same as those expressed when the men actually occupied the hand mill jobs. But on this occasion there was a change of emphasis which we will discuss in the following section.

One interesting point was that a larger proportion claimed to have been satisfied than was shown in the sample before the transfers. Table 16 illustrates the difference in overall job satisfactions as expressed in the interviews before the changeover, and the post-transfer responses where the interviewees expressed their satisfaction in retrospect.

Table 16. Job satisfaction replies

	Satisfied *Per cent.*	*Dissatisfied* *Per cent.*	*Undecided* *Per cent.*
Actual (pre-transfer) ($N = 47$)	70	19	11
Retrospective (post-transfer) ($N = 51$)	92	6	2

What might account for this difference in the recall of job satisfaction? It could be a *golden age* effect through forgetting of unfavourable aspects, demonstrating a general tendency for those parts of experience which represent a negative significance to be forgotten more rapidly than are positive stimuli. Or this could be a type of gradient effect—where the individual experiences the negative incentives of a situation as being less significant or powerful, the further away from it he is.

This was not so for all the negative features of the old jobs. When the specific likes or dislikes of the retrospective answers are examined in Figure 12 we see that although there was some reduction of factors recalled, the major disincentive—physical work load—received a heavier weight of opinion

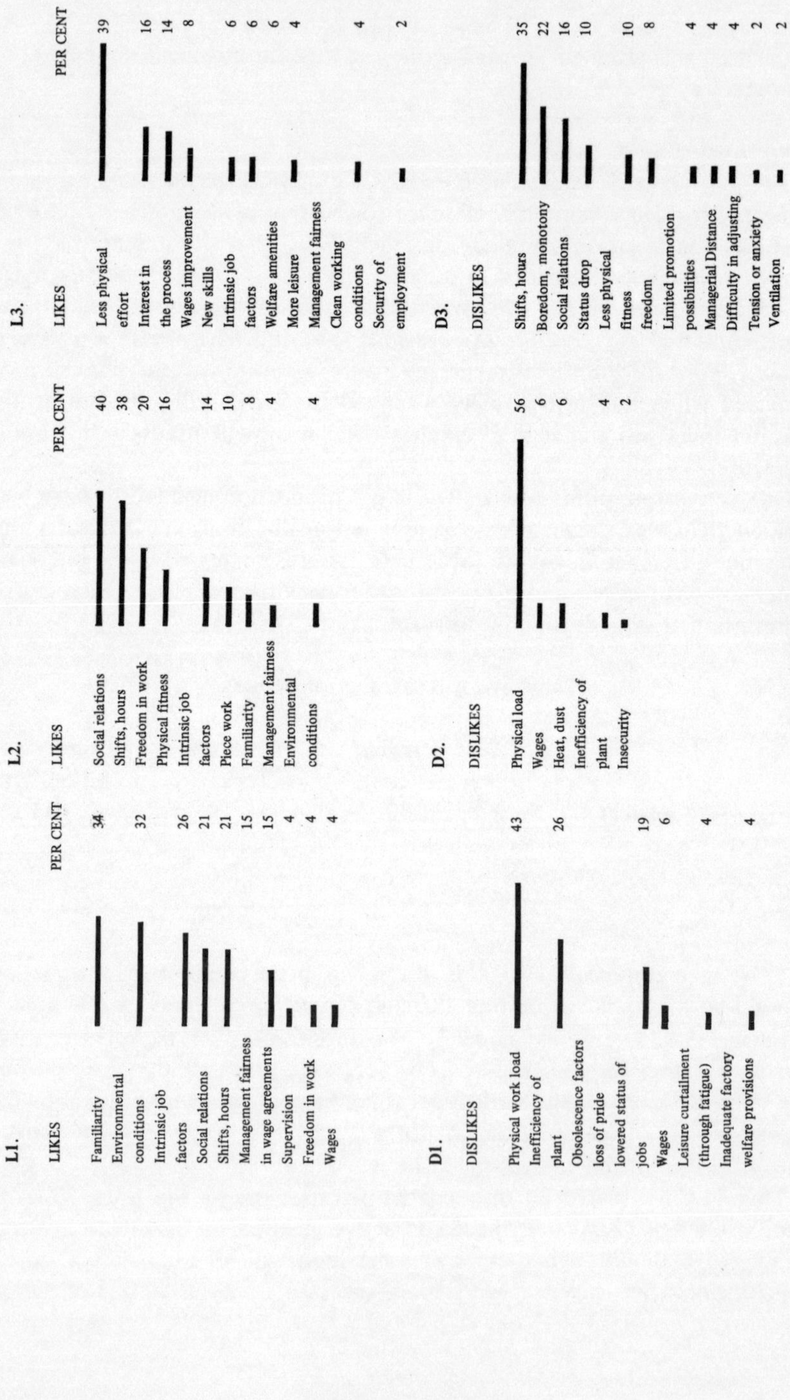

Figure 12. Likes and Dislikes

in retrospect. Some positive aspects also attracted more comment in the retrospective replies (L2) than before the transfers. As the bar diagram illustrates, the topics referred to by the interviewees at both points of time (L1 and L2) were largely constant, and seven out of eleven factors were mentioned on both occasions; however, the rank ordering of these is markedly different. The factors given most weight in retrospect were the converse of present dissatisfactions (see D3 in Figure 12), and the four major retrospective areas of satisfaction are present among the first half-dozen dissatisfiers of the new job.

We have a good deal of evidence on the complementarity of the retrospective and present attitudes. The responses, L2 and D2, of the recall question in general follow the reverse pattern of L3 and D3.

The majority of the current dislikes took the form of comparisons. With the exception of the boredom effects, the other negative factors in the new jobs were perceived as deprivations in the light of experience of the previous environment. Shift arrangements, in particular, and social relations are the outstanding examples of this type of deprivation. There was also mention of less personal freedom in the immediate work environment. This resulted partly from the fixed speed of the machine system, which was contrasted with the variation of pace in the hand mills, and partly from the perceived supervisory controls which were compared with the former virtual absence of supervision.

Other similar comparisons occurred in the opinions expressed on the interviewees' lowered condition of physical fitness, the increase in managerial distance from employees and fewer promotional opportunities. Although this last grievance derived from the placement policy referred to in Chapter 8, it touched on a fundamental change in the control and administration of promotions which had formerly been in the hands of the employees through their local union branch. We shall discuss this in the analysis of the primary group in the new plant.

Conclusion

If we assumed that human adjustment should necessarily follow innovation, we could fall into the error of thinking that adjustment was desirable without taking into account the objective elements in the change. It would be unrealistic to expect the employee to accept innovation without reference to its meaning for his work or leisure life or to the ways in which the changes are implemented: these could well be ineffective or ill-formulated for their human effects.

The cliché of attributing *resistance to change* to industrial employees will not suffice as a let-out in this discussion. Quite definitely, irrational resistance to the notion of technological innovation itself is not encountered among the empirical findings on automation.[3] There may be resistance to certain effects of the innovation, for example in relation to a perceived threat to established promotion arrangements of the sort which Mann and Hoffmann described in a study of attitudes towards change in a power station.[4] On the other hand, resistance to change in attitudes, the persistence of expectations for certain desirable inherent qualities in the environment, is not necessarily bad or the new environment always good or better in all respects, to the old.

For these reasons the solutions to the human problems of innovation are only partially found in programmes for improved downward communication which aim at attitude-change among employees. Although communication procedures seem to be particularly necessary in advance of the introduction of new production methods, yet it is more important that they should form part of a general plan for organizational change which includes a reformulation of job and social behaviour at all levels of the hierarchy. Specifically, this means that there must be redefinition of job roles and duties throughout the organization in terms of the interpersonal relationships which are required both in work teams and vertically at the points of contact between the hierarchical levels. If this can be done searchingly and, in sufficient detail, with the necessary expertise, then attitude-change itself will occur in a positive direction almost as a matter of course: because the appropriate behaviour will have been considered, planned, and introduced during the period of preparation.

Management and the unions as yet have not had the expertise available, even if they had the inclination, to deal with change in its less obvious human aspects; and we are still somewhere along the way to accumulating the empirical information upon which such expertise would be based.

Summary

We have seen how the process of individual adjustment began in the period before innovation with anticipatory categorizing of the likely outcomes. Job and social learning followed in the initial months, although the latter took place over a longer time interval, and in Chapter 11 we shall discuss the particular problems of social learning.

At some point after initial learning it became possible for the individual to make an evaluation of the advantages or disadvantages of his present job. The sequence of adjustment was now completed. The interviewees showed a readiness to accept the change in general, but difficulties did arise from job content and the immediate environment. Responses relating to monotony in jobs, subjective feelings of fatigue and 'brain work' were discussed. Over half the interviewees said they were more tired in their present jobs and for a large proportion boredom was associated with feelings of fatigue. A motivational factor was suggested. In many instances jobs did not possess intrinsic incentive value, and added to this was the individual's very limited knowledge and understanding of technical aspects. The job had the attractions of higher wages and better working conditions but there were further negative effects in the social relations surrounding them. A retrospective question threw more light on the positive evaluations given to the former jobs in this respect. In the following three chapters we will examine the social dimension in the new plant: the work group, supervisory changes and managerial policies. These were features of the total social organization, treated separately for the purpose of analysis.

REFERENCES

1. Jerome S. Bruner, Jacqueline J. Goodnow and George A. Austin, *A Study of Thinking,* New York Science Editions, 1967. Here the process of categorizing is described as achieving reduction in the complexity of one's environment, identifying, ordering and relating classes of evidence. The authors emphasize the anticipatory nature of much of our categorizing where meaning and identity are attached to events before they occur by the utilization of prior signs.
2. Charles R. Walker, *Toward the Automatic Factory,* Yale University Press, New Haven, 1957.
3. See Avner Hovne, 'Some Social Implications of Automation', *Impact,* **15,** No. 1 (1965), pp. 5-25.
4. Floyd C. Mann and L. Richard Hoffmann, *Automation and the Worker,* Henry Holt, New York, 1960, p. 95. See also Alain Touraine, Claude Durand, Daniel Pecaut and Alfred Willener, for a general discussion of resistance and acceptance of change in Chapter 2, 'Change and the Defensive Reaction' in *Workers' Attitudes to Technical Change,* Organization for Economic Co-operation and Development, Paris, 1965.

Chapter 10

Factory and Community: After innovation

In the preceding chapters the individual's experience of changes in the factory environment has been analysed. However, dissatisfactions with the continuous shifts were expressed in terms of effects external to the factory. It was evident that this result could be understood only by referring to the regularities of behaviour established in the local community. More generally, the interviewees' perceptions of changes in their leisure activities might reveal how the innovation had affected community structure and its relationship to behaviour at work.

There were two major factors which produced a separation in social ties of the kinds which were described in Chapter 4. These were, briefly, the recruitment of employees from a number of scattered urban centres and the new shift arrangements. In addition, the siting of the factory in open fields some miles from any of the townships where employees lived contributed to this separation. The men travelled by car or by public transport instead of the previous short walk from home to factory, although the increase in the journey to work was not an important factor in itself. But there were indirect social consequences. It had been customary for production teams to meet on their way to work or to return together: 'it was a regular thing in the old works to have a glass of beer on the way home'. The interviewees felt the loss of the influence of a single close community around the factory, as this ancillary worker mentioned 'people live in a scattered area, they want to get home as soon as the shift is over' or as a remark by another operative hinted, 'we were a gang on our shift, we'd call in and have a drink together. Here you are an individual among people from other areas'.

A question was asked on meetings with work mates in leisure time. The replies were almost all in the direction of fewer outside contacts, as shown in Table 17.

Three men who saw more of their work mates, frequently visited the new Works Social Club alongside the factory. Four others who thought that contacts were about the same still managed to meet their *former* work colleagues in their local Workingmens' Club. The remainder saw less of their work mates including eleven who had never met anyone from their workshop or production line in leisure hours. As a chief reason for this change the interviewees

Table 17. Do you see your work mates outside work, more or less often or about the same as in the old plant? (*N* = 48)

Less often		*About the same*		*More often*	
No.	*Per cent.*	*No.*	*Per cent*	*No.*	*Per cent.*
41	86	4	8	3	6

mentioned the distance between their homes. Shift rotas were cited by those who interpreted 'work mates' as meaning others transferred from the old plant. Thus, the distribution of reasons for not meeting outside was as follows:

Table 18. Reasons given for seeing less of work mates outside the factory (*N* = 41)

	Number mentioning this reason
Distance between employees' homes	27
Different shift rotas	15
Other personal reasons	2

The community connection had undergone a very great dilution from an extremely intense set of relationships to the almost complete disappearance of ties with the factory. This interviewee transmitted something of the closeness of the cross-ties before the changeover which might seem almost oppressive to an outsider:

'All the people I worked with lived in the same town and we were in contact after work. My mate with me lived next door and we all lived around one another. You would go to one another's houses'.

Another man explained the change in this way: 'before we were drawn from within a square mile of the works, now we are drawn from four to twelve miles apart'. All this pointed to the conclusion that reinforcement from leisure activities for the primary group inside the factory had not continued to anything like a similar degree. There was a cessation of the neighbourhood influences which in so many instances counterpoised the authority hierarchy in the old plant. In Chapter 4 it was seen how local associations affected both the administration of managerial authority and the development of norms in the primary groups, i.e. through the incorporation of neighbourhood values in the normative behaviour of work teams.

Shift work and the community

In Chapter 8 unfavourable opinions of continuous shifts were shown to be related to their interference with habitual and collective activities of the weekend. It was demonstrated how a small minority of the men preferred the Continental week to the weekly shifts because they now had more time for their hobbies, home interests and families. Wives' opinions were almost entirely favourable towards the new shifts for similar reasons. Possibly a new pattern was evolving, where husbands shared more leisure time with their wives. Round the clock shift working did not appear to produce less domestic contentment, as has been feared.[1]

The conclusion that it is not domestic life so much as community activity which suffers most from shift working was emphasized in Mann and Hoffmann's study *Automation and the Worker*: here the analysis showed that friendship relations in the community were 'the most sensitive to the deprivations of shift work'.[2] Wallace Bloom expressed this general problem of community effects which continuous shifts may produce when he gave examples of shift workers 'forced to live in a different time sequence from that of their community and the many people to whom they were intimately related'.[3] Shifts may bring about a pattern of leisure which is concentrated more on the home and on the nuclear family unit. The arrangement of working hours can thus be seen as an influence working against the commitment of the worker to various social groups in his behaviour field and a limitation on what Lockwood has called the occupational culture: 'These primary groups of work mates not only provide the elementary units of more extensive class loyalties but work associations also carry over into leisure activities, so that workers in these industries usually participate in what are called "occupational communities". Work mates are normally leisure-time companions, often neighbours, and not infrequently kinsmen'.[4] It can be concluded that we have to look beyond behaviour in the factory itself for explanation of the dissatisfactions with shift arrangements and to open up the frame of reference to analysis of the current expectations about leisure time in the worker's own community and neighbourhood, the nature of the community and its values.

Wives and the transfers

Thirty wives were interviewed who were already married during their husbands' employment at the old plant. The wives were asked their opinions of the general consequences of the transfers. Twenty of these found a marked or definite change in their lives, only ten stating there was little or no difference.

Table 19. Conditions since the transfer: wives' general evaluation ($N=30$)

	No.	*Per cent.*
Very much better	17	57
Better	4	13
A little better	5	17
Neither better nor worse (no change)	2	7
A little worse	1	3
Worse	1	3
	30	100

Evaluating the direction of the change—towards improvement or otherwise for themselves or their husbands—fifty-seven per cent. answered on a six point scale in the most favourable category of 'very much better'. In all, twenty-six wives considered that the transfers had produced improvement in some degree, only two saying that there was 'no change' and a further two considered that their lives now were 'worse' or 'a little worse'.

The most frequent reason given (by almost a third) *why* the present situation was better, was that the husbands now spent more of their free time around the home: 'I see more of him' was a characteristic reply. The same number of respondents also mentioned that there was less washing because of cleaner working conditions and several thought that things were now 'easier for the wives'. Seven of them drew attention to the improved wages of the plant and the lighter work required in their husbands' jobs. When asked later if there had been any changes for the worse, several talked about weekend working and the loss of opportunities for attending social activities or church; one wife missed 'going to Saturday cricket together'. But just under half the wives thought there had been no changes at all for the worse.

To conclude, the majority of wives found that the transfers to the new plant, its shift system, wages, and light and cleaner job duties, had brought an improvement into their daily lives. The replies revealed greater emphasis or dependence on the nuclear family in the allocation of leisure time, but particularly for the husband, a discontinuance of membership of social groups and participation in collective activities away from the home.

Earlier, I described the influences which operated in the interviewees' adjustment to the new factory and this brought into consideration the interests and customary behaviour of the men outside of work. It was seen how, before the changeover, there was a link-up between work and community in associations of a mutually strengthening nature. Since the transfers, fragmentation had occurred because of the distribution of employees' residence over a wide geographical area of seven or eight urban communities as well as from the effect of shift working on leisure activities. In place of these activities there was now the possibility of meeting during leisure time in the new social club attached to the plant but only in five interviews was the club mentioned.

It can be concluded that there had occurred a separation of behaviour *inside* from that *outside* the factory. At the same time, those activities were fewer where the individual's interests and identity in leisure and work merged. The rhythms of community and factory were no longer one, and the social distance between worker and worker and between worker and manager was greater, not solely because the new plant was several times larger in size than the old. Attached to the previous jobs were a number of values developed through family and neighbourhood, and the interviews showed the reduction in these more extensive social meanings of their jobs for the workers who were displaced.

Summary

The features producing a separation in social ties between factory and community were described. There was less influence on factory behaviour from neighbourhood families and associations such as existed in the period before innovation. Moreover, a large majority of workers had little or no opportunity for meeting work mates in leisure hours, and shift work interfered with the traditional collective activities of the weekend. There was now a severance of the connection between work and behaviour in the community. But interviews with employees wives' showed that the outcome was a positive one for domestic life, bringing greater emphasis on the home in the use of leisure time.

REFERENCES

1. Leon Bagrit, *The Age of Automation,* Harmondsworth Pelican Books, 1965, p. 81.
2. F. C. Mann and L. R. Hoffmann, *Automation and the Worker,* Henry Holt, New York, 1960.
3. W. Bloom, 'Shift Work and Human Efficiency' in E. A. Fleishman (Ed.) *Studies in Personnel and Industrial Psychology,* Dorsey Press, Illinois, 1961.
4. David Lockwood, 'Sources of Variations in Working Class Images of Society', *Sociological Review,* **14,** No. 2 (1966).

Chapter 11

The Social Group after the Changeover

In order to chart the structure and forms of the primary group in the new factory and to define the differences in interpersonal behaviour which have evolved since the changeover, several sources of data were used. Two of these consisted of interview responses: namely those of the free-answer interviews where opinions were expressed spontaneously after an initial general question, and in addition, the distributions of replies to direct questions on aspects of behaviour in the work group. A further source was an analysis of work-group communication networks after the activities of four production teams were recorded over a period of several weeks. The point of the latter was that it provided a measure of behaviour to which it might be possible to relate the opinions and attitudes of employees. In brief, what was the new pattern of activity or communication between operatives? If there were negative attitudes, to what parts of the social environment did they refer?

Before the transfers social relations in the factory were fourth of the nine satisfiers mentioned. In the new plant a reversal of attitudes was demonstrated (and discussed in Chapter 8) from favourable to unfavourable, bringing this topic to third place in the rank order of dislikes. Again, this tendency was emphasized when interviewees were asked after the changeover to list their likes or dislikes as they *recalled* them in their old jobs: social relations was the positive characteristic most often mentioned, with forty per cent. of the men giving it this retrospective primacy. In Table 20 we can compare the six major features of jobs or their environment which were the subject of favourable opinions in the old factory, with the frequency order of the first six as they were recalled.

So the satisfier effect of social relations in the old factory appeared most strongly in the answers to the retrospective question. Although there might have been some recall (golden age) distortion, we have already noticed that interviewees placed more importance on some aspects of the previous job by contrast to perceived current deprivations. It could be concluded that what they previously assumed as usual or customary came into prominence by the

Figure 13. An Electrolytic Tinning Line

effect of innovation in making certain social forms or advantages inaccessible.

Table 20. The previous job: Actual and retrospective satisfactions

Likes (actually expressed) in the previous job	*Likes (as recalled) in the previous job*
Familiarity of work place	Social relations
Environmental conditions	Shifts, hours
Intrinsic job factors	Freedom in work
Social relations	Physical fitness
Shifts, hours	Intrinsic job factors
Management fairness	Piece work

It soon became clear that there was not merely a modification in the behaviour described in Chapters 2 and 3 but a substitution of major elements in the technological and structural organization of the social group at work. For the men who were interviewed it was not a matter of varying any of the established routines to different tolerances but of learning behaviour to suit entirely different criteria of job requirements and relationships. These were more than changes in variables of the social system—they were larger-scale changes of environmental parameters which made inappropriate the norms or standards which the individual had previously acquired. Using the terms of analogy to physiological or engineering systems, it was not a matter of moving from one state of activity to another, with the values of the system maintained within stable limits: not the type of change of which behaviour itself consists, but a qualitative change from one entire field and structure of behaviour to another.[1]

How was the result of this transformation expressed in social roles with work mates or staff members? It was seen there was no social field of the complexity described in the early chapters. In fact, the recurring themes of the interviews concerned not so much the primary groups of the automated line as those of the old factory. In comparison with the diverse functions of the latter, the present environment seemed featureless.

After the interview question on job satisfaction, another opportunity for spontaneous comment was given by the question: *If someone were to ask you, what is the main difference between working at the old plant and working here, what would you tell them?* While improved physical working conditions were mentioned by sixty per cent. of the interviewees, the second main feature of the changeover was a negative one: fifty-two per cent. drew attention to what they called the difference in the social 'atmosphere', referring in unfavourable terms to present relationships with work mates. A large body of opinion drew attention to the 'family spirit' of the former primary groups, in comparison with a situation where interviewees perceived they were 'individuals' with a few or superficial contacts with other employees in or outside of work.

A question was used which sought information on whether interviewees were members of a team or worked on their own. Table 21 presents a clear result.

Table 21. Are you a member of a team or do you work on your own? ($N = 51$)

Member of Team		*Solitary workers*	
No.	*Per cent.*	*No.*	*Per cent.*
38	75	13	25

The replies on membership of a 'team' showed that seventy-five per cent. of those interviewed belonged to teams of six to ten members or of working pairs in the case of maintenance staff, with the remaining twenty-five per cent. working on their own. The question did not give further leads and it did not differentiate between employees in terms of their relative isolation. For example, the production operatives were all nominal members of teams although this concealed differences in locational separation and in the interpersonal contacts required of individuals.

As a further check on these factors, employees were asked whether they talked to others more or less than in their old jobs. The replies gave this distribution:

Table 22. Do you talk more or less or about the same as in your previous job? ($N = 48$)

Less		*No difference*		*More*	
No.	*Per cent.*	*No.*	*Per cent.*	*No.*	*Per cent.*
19	40	15	31	14	29

The tendency towards reduced incidence of conversation with others was associated both with the physical distance between operative positions and with the vigilance requirements of line jobs: 'If you are concentrating you've got to listen for defects, you can't converse so much because something can go wrong'—this was one of several references by inspectors and others to auditory cues on the line. Managerial policy was also in evidence; one feeder pointed out that 'sometimes if they (managers) see you talking they don't like it. It depends on the way the tonnage is going. They don't like you in groups talking, you can't be in groups anyway because each man has his position'. Another, explaining that there were fewer opportunities to talk than formerly, mentioned that there was 'more discipline, they say "Mr. So-and-So is coming" and you try to look busy'.

However, the men who had been transferred from production to ancillary or maintenance jobs clearly had more opportunities for talking: 'I can talk very often—if there is no-one watching'. Some had difficulty in comparing two situations of very great qualitative differences. In this respect the real issue for them was not social conversation but the relationship of their job roles to those of others. It may be significant that none of the interviewees called attention spontaneously to opportunities for conversation although they discussed a great many other social effects of the innovation.

Analysis of social interaction

Interpersonal behaviour in work teams was observed in order to establish whether there was a pattern which could be quantified and compared with the previous one. All contacts initiated and received between each of the nine operatives were recorded in an interaction matrix of 72 (9 x 8) cells, illustrated on the right of Figure 14. One contact represented a communication initiated by one member of the team and received by at least one other member. Although on this production line a contact was usually a communication, i.e. involving transmission of a message, it might also be an occasion of shared activity between team members (as was especially the case in the pre-changeover team).[2] Observation was made of four teams in electrolytic tinning for periods of 25 hours in each. The interaction matrix revealed the degree of usage of the seventy-two possible communication paths between the operatives, any one cell representing the sum of initiating and receiving contacts between a pair of operatives.

Figure 14 shows the communication frequencies of each of the four production teams treated separately. The *unbroken* lines represent the curves of formal communication, i.e. as required by the technical duties of the operatives. The *broken* lines are contacts between the men which were clearly not required by the job operations. This distribution of frequencies gave an outline of a consistent pattern in the primary group, the results demonstrating a low interaction rate for the majority of cells. The graphs thus show a J-shaped curve of extreme skewness—most cells or paths were relatively unused and the long tail of the distributution extends towards the right, where the frequencies are larger.

It is easily seen that the communication in the four production teams yielded very similar curves for each. There was no significant difference

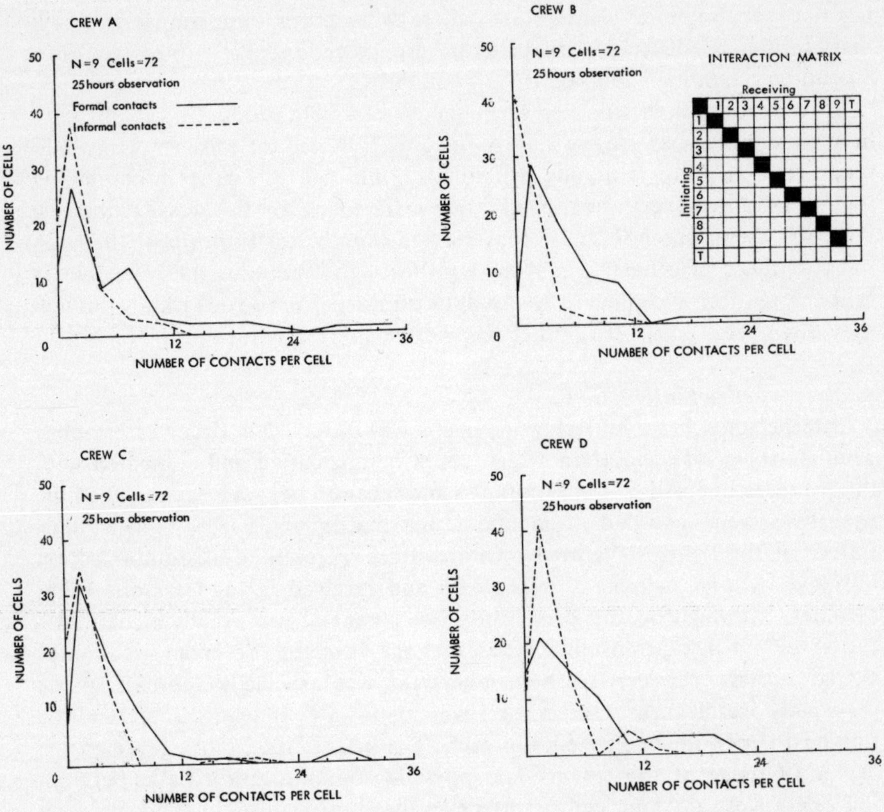

Figure 14. Contact Frequencies

between the modal frequencies. It could be deduced from this that the low frequencies of contacts which were classified as job-required, seemed to have been insufficient to provide momentum for elaboration of informal activities. However, features of the technical environment, e.g. machine noise and locational distance, may have tended to make such an elaboration more difficult to achieve. We can assume on the evidence of Chapter 2 that in the hand mill crew the curve would have shown the opposite distribution, with a strong skew to the upper frequencies.

Attitudes and behavioural roles

As we have seen in their replies to the *main difference* question, just over half the men described the change in 'atmosphere', emphasizing that the 'friendliness' of the work place before the transfers was now lost. In the former work groups the security experienced by the individual in a clearly established set of social roles extended to the informal roles among the operative positions and across the grade hierarchy in the work teams. But the known routines included even the idiosyncratic behaviour of individuals. The men in the new situation were explicit on these advantages which they had enjoyed in the old factory and which they now expressed in very similar terms to the interviews before the transfers:

'You knew everybody and you knew their feelings, you had worked so long with them you were practically brought up with them. You knew you would be in a crew where you knew every one of them, individual, and you knew his troubles as well. You knew the man you could joke with and the one who you couldn't'.
'Now when I think about the old works, you seemed to get on with your work with one or the other cracking a joke, but still you'd get on with your work. More like just your circle of friends, working together and each one knowing the other'.

Community influences also appeared in those replies where the interviewees described the change in 'atmosphere', giving examples of the less varied relationships between work mates in the new factory. They emphasized that since the transfers there were only surface relationships, compared with acquaintanceship with individuals in the multiple roles of family or neighbourhood:

'There is a difference here in the atmosphere. Before you more or less knew everybody by family. You knew their brothers, sisters, wives and their children. But here, well, you just know the man you are working with'.

These last excerpts highlight the cumulative effects in the old plant of long service, kinship relations, and the single community background of the men. Inevitably, at the changeover, these influences were dispersed among the much larger population drawn from a wide geographical area which spanned a number of different urban communities, as was described in the last chapter.

If the effects inside the factory were partially derived from the severing of the community connection, they were assisted by the separation of members of the production teams which was not only locational but was induced also by differences in behaviour. These differences had their expression in the interviews:

'There seemed to be more unity in the old factory than in a place like this where you are so scattered. All the shift you would work as one, more like mates, then'.
'Perhaps it's that we are not banding together here as in the old place, there are more machines and more concentration . . . the good will towards one another was greater than it is in this works'.

These conditions had little to do with the length of time employees had been in the new production lines as this remark illustrated: 'you work with them and you don't know them, we are strangers after three years'. The team had settled into a much less close, even a competitive, level of functioning where the men could not 'confide in others'. A fitter explained that 'there was more comradeship in the old place, here they like to see you making mistakes', and a production operative:

'If you were learning they'd give you their tongs they'd be willing for you to have a go. Here they're like as if you are going to take their jobs off them. They used to work as a crew. They don't do it here'.

This reflected a pattern which would perhaps be more recognizable among professional groups of high individual aspirations and mobility. It was also not improbable that what created difficulties for these interviewees may have been welcomed by other workers less tied to the traditional ways. There was a small clue of this in an interview with a boiler-room attendant, one of the younger members of the sample, who expressed an interesting opinion:

'I have heard a lot say that the "atmosphere" here is not so good but don't forget I am younger and that old way used to get me a bit. You would meet a bloke who would know more relations of mine than I would know myself'.

Future research will no doubt test the possibility of alternative patterns of job rewards acceptable to more recent entrants to the automated industry.

There were large contrasts in the qualitative nature of behaviour in the factory. A loss of cohesive quality in production teams was likely to derive from the absence of individual choice procedures. As well as opting into a team it had been just as important that a man could withdraw from one if he wanted. On the new production line it was neither possible to choose a team nor escape from one if personal difficulties arose. There was already something of a folklore about what happened where there were animosities between operatives. The following is an example of other similar narratives which were in currency.

On one line the relief operative (piler) who took over the job of feeder, seemed unable to weld one coil to another at the entry point to the line without causing a stoppage. Normally the preceding coil had enough slack for the continuous movement to be uninterrupted. But during the forty minutes of relief when three or four welds would be made, the line frequently stopped. As this continued over several shifts the remainder of the team began to jeer and to applaud each time a weld was made. Eventually the relief feeder accused the shearman of pressing a stop-button at the moment of the weld and counter-accusations were exchanged among different members of the team. Some time after this incident the shearman discovered an electrolyte attendant pressing the button.

There was no evidence in the interviews of the prescribed interpersonal roles which were characteristic of the situation the men had left behind. If, in the electrolytic tinning production team, there had not developed high rates of interaction or voluntary job sharing through the grade hierarchy, neither were there strong social controls similar to those of the old factory. As a corollary to this the individual was also without the aid of guidance and assistance such as was supplied in the primary groups before the transfers. The men recalled that in the old production crews the individual could depend on other members of his crew: there were secure expectations that work mates would guard his interests. Clearly, the mutual 'cover' which was provided by the hand mill group had not been maintained. The implications were shown in the references to a number of unfulfilled cover expectations ranging from straightforward assistance in job tasks to more subtle forms of support:

'It's a case of knowing that if anything did go wrong, they would cover you up (in the old plant) but it is very doubtful they would do that here. It's a job to explain. If you were down on your job, your mate could help you, you can't here'.

'The atmosphere is totally different to what it was before. Before we were a team, a unit, now we are more individuals. In the other job you used to make your own fun and you used to depend on others more. You seem to drift away from the men more up here than before, every man to be for himself'.

References to the old norms

Thus the interviewees continued to use previous norms as a standard in assessing present relationships: the result for the latter was pejorative as the following examples show. The men utilized the norms of harmony, interdependence and mutual help as their source for comparisons:

'There did seem to be a different feeling then in this place. Whether we depended on one another because we were all pulling our weight, everybody was helping one another, really, and if anybody was down I found it myself you would always have someone to give you a hand. There was the comradeship, there had to be, it was no good one saying "I will do this my way", we had to be as one, really, that is what it amounted to'.

Presumably, whatever might be the behaviour of work groups on the automated line they could be expected to develop different forms from those of the hand mill crew. In the latter, direct modifications of pace and sequence of activities was possible by the intervention of the crew members, who had control over technical operations. Obviously the norms governing this behaviour could no longer apply because both the standards of performance and the sequence of operative roles were determined in the new production line to a far greater degree by the requirements of the machine system. Moreover, the control exercised by the operating crew in the hand mills was now shared with supervisors, maintenance staff and management who used technical criteria beyond the range of skills possessed by the operatives.

In summary, the work group in the new factory did not exhibit the development of closed-loop behaviour and control either over technical factors or interpersonal relationships. From the point of view of the interviewees, the difference mainly appeared in the absence of comparable reciprocal activities. In the current situation norms of reciprocity existed in much lesser intensity and if we were to assume a high-low continuum for supportive behaviour according to its frequency and variety, the present group would fall well into the lower end. It is true that the aims of the pre-automated crew were partly to relieve negative pressures on its members, but there was need for a similar function in the present environment, although towards a different set of negative factors. The interviewees effectively expressed just this. However, it should be recognized that support may now have been sought to a degree which was the result particularly of conditions of stress associated with innovation effects rather than with automated technology in general.

Union differences

We may conclude that a set of activities prevalent before the changeover had been both modified and truncated. As well as the changes in the immediate organization of the work groups there were other more general influences on social behaviour but tending towards similar effects: interviewees mentioned more numerous demarcations in union membership between occupational strata or sections of the plant. These tended to add a further behavioural separation. A crane driver said that he preferred the 'harmony' of the old plant: 'here it's different altogether, the unions are one against the other, the men fighting against each other, the harmony is not the same'. The effect on men whose experience had been of one-union membership appeared in several interviews, with the emphasis which is illustrated in tnese examples from the spontaneous comment of two ancillary workers:

'There is a difference between the men. We were all semi-skilled in the mills. Here we have skilled men, tradesmen. There seems to be that inferiority feeling between some of the mill men and the tradesmen here. There is that difference. You don't seem to be growing with them as we did in the old works. There are things I'd like to say which I can't say'.
'If there is any problem among ancillary workers on the line the production workers don't show any interest, and the other way round. And you see the unity is lost. In the mills the problems of a back-roller on No. 6. mill concerned the first helper on No. 1'.

These union differences were examined in relation to attendance at the local meetings. Thirty-six of the interviewees were asked to compare their present rate of attendance with that in the previous union branch. The results are shown in Table 23.

Table 23. Do you go to union meetings more or less than before, or about the same number of times? ($N = 36$)

Less		*About the same*		*Not at all*	
No.	*Per cent.*	*No.*	*Per cent.*	*No.*	*Per cent.*
24	67	5	14	7	19

Although fourteen per cent. said that their attendance was unchanged, sixty-seven per cent. said they attended less than previously and a further nineteen per cent. who used to attend occasionally in the old plant said they did not now attend at all. Thus, although some interviewees were not asked this question, a strong majority claimed little or no attendance at meetings. A closer look can be taken at the reasons given for the decreased attendance at meetings. Thirty-one of the men were asked *why* they attended less and the following distribution was found.

Table 24. Reasons for non-attendance or decreased attendance ($N = 31$)

	Number of interviewees giving this reason
New shift arrangements	18
Inconvenient journey to meetings	14
Differences in union activities	6
Other leisure interests	4
General apathy among men	1

Most interviewees gave one reason for non-attendance. The new shift arrangements were stated as the main obstacle (by eighteen) and the geographical scatter of the employees' place of residence in separate urban communities was mentioned by fourteen of the men as creating difficulties for travel to the union meeting place. As we saw earlier, branch meetings were one of a number of leisure group affiliations which were interrupted or curtailed because shift arrangements reduced the total of employees available for attendance. Before the transfers the meetings were held on Sundays when men from all shifts were free. It will also be recalled that meetings used to be held in the community where most of the employees lived. Now, although the union meeting place was chosen for its equidistance, roughly, from several urban communities, this did not really appear to minimize the inconvenience of a longer journey to the meetings.

Six interviewees mentioned that the content of meetings did not touch on matters of personal interest or that there was less possibility of redress for any complaint at the branch level. In fact, it was not possible to give a detailed treatment of union branch activities in the automated plant because they did not exist on the previous scale in workshop matters. Formerly the local union had used its organization to achieve the immediate demands of work teams and individuals. Now the focus of union businesss tended to move away from activities of the work group and upward from the local branch unit to the level of divisional organization and of general agreements in the industry. Since the branch did not continue to provide its members with immediate rewards at the level of the team, we might speculate that this had some effect on participation in union affairs as expressed in attendance at branch meetings. It will be remembered that in the old plant it was in a man's personal interest to attend because of the presentation of promotion and vacancy claims. One worker recalled how he used to attend more often before the transfers because at the meetings 'there were always little jobs cropping up, promotion quibbles . . . the superintendent moves the men now, not the branch'.

While the union officials in the plant continued to operate the seniority principle for promotion, they did so only to the extent of presenting to management the names of a number of men who would be eligible when a vacancy occurred. The branch itself did not implement promotional moves as described in Chapter 3. Neither did the branch have power to allow team transfers and demotions. In a sense too, this change in activities had affected the individual's participation in the social structure of the work team: in the old plant he could assist in the promotional procedures, intervening personally to state his own claim or contest those of others and he influenced the formation of the production crews through the team choice device.

Motivation in the primary group

The adjustment problem in social attitudes and behaviour could be summarized as follows: essentially, the men had moved from jobs where incentives for task performance were found in the immediate primary group. Motivation from this source was likely to be effective because the occasions for reward from the technical and social roles of operatives were extremely frequent: they were the variation of task sequence, exchanges of jobs and the enlargement of task area, recognition from others and the consequent and longer-term opportunities for promotion.

In contrast, the automated plant offered fewer rewards to the men which were intrinsic to the job or the work group; individuals participated less both in the production operations, and in reciprocal social behaviour. Instead, incentives were found to lie in the interesting technical content of the machine system and in improved wages. There were, however, other negative incentives or dissatisfiers and while the interviewees, in the majority, expressed an overall satisfaction or acceptance of their situation, this general attitude co-existed with problems of ambiguity in the norms of appropriate behaviour in the factory. The individual involved in the changeover was required to make a redefinition of his role and its rewards but these workers were unwilling to give up the values of the old work crew; what we might wish to consider quite carefully is the question of whether these really had to be discarded.

Conclusion

Ways have been described in which social relations were a source of grievance to interviewees. Very often they raised the topic of the differences between behaviour in the former production teams and that of the automated plant, generally to illustrate their unfavourable attitudes towards the latter. In the description, earlier on, of the individual's job in the old factory, it was demonstrated that the content of the job and group behaviour were closely associated variables. In the new production line social learning was important in the individual's adjustment and emerged as a major source of difficulty. Again, the tedium experienced in monitoring tasks was related to the structure of job contacts and the isolation of operatives.

If the social dimension is to be included in planning for innovation it will be necessary to make an assessment of alternative outcomes, setting out explicitly the interrelationships (and interchanges) of each operative's tasks as part of design forecasting, and influencing the technical organization to be adopted. Even though the models constructed could only be tentative, what is really required is that this dimension should be brought into focus for design purposes in its detailed and measurable aspects.

It may have to be decided which of the forms of social behaviour might be prevented from varying too much in order to give the individual reliable points of reference to assist him in coping with innovation consequences. As constants in the system these would reduce the jarring impact of change occurring in other sectors of the total environment. A structural variable which might have been maintained at constant levels was the rate of interaction between individuals. No matter that this seems to be a relatively narrow segment of structure, it has meaning for the formation of attitudes which in turn generate quality and quantity of behaviour. In the final chapter there will be a speculative discussion on primary group variables which might be included in planning.

Summary

There was a reversal of attitudes towards social relations in the new factory compared with the old. The individual's experience of uncertainty in relating his role to that of others was described. Behaviour in four production teams was subjected to analysis. One segment of the primary group structure—its frequencies of interpersonal contact—was abstracted for intensive study. It was shown that there was a consistent pattern of sparse activity between team members. Task-required contacts were not sufficient in frequency to produce an elaboration of interpersonal activity with effects for job content as described in Chapter 3. Reference was made to differences in union organization and attendance, and their significance was further considered in relation to changes in the primary group.

Interviewees' opinions revealed that their established expectations and behavioural standards were inappropriate to the present situation. Less commitment was required to team exchanges and relationships but less support was afforded, and the latter might have been especially useful to the individual in conditions of large changes in the job environment. Following on from this assumption it could be questioned whether the technical reorganization should have been permitted to disrupt existing group behaviour. This raises the general question of whether such forms could be included for design forecasting in explicit and measurable terms. If this could be done a significant contribution to planning might be the formulation of an optimal structure for the primary group.

REFERENCES

1. W. Ross Ashby, *Design for a Brain*, Chapman & Hall, 2nd edn, 1960, p. 73. Here the distinction is made between change of a variable and change of a parameter. The former is considered to be the kind of change-value of which behaviour itself consists and which is an inherent property of dynamic systems. The latter is the

more drastic case of change to a completely different behaviour pattern – 'from one behaviour to another'.

2. The method of observation and coding of communications was discussed in J. K. Chadwick-Jones, C. Sheppard and A. Carceller, 'Observational Recording of Social "Load" Among Operating Crews in Continuous-process Industry', *Proceedings of the Third International Congress on Ergonomics,* Birmingham, 1967, and J. K. Chadwick-Jones, 'The Measurement of Social Communication in Industrial Work Groups', *Bulletin of the British Psychological Society,* 20, January, 1967.
A contact was defined as occurring when one person talked or signalled to another and the latter made a response (oral or by signal) directed to the person initiating the contact or otherwise acted on it, for example, in following an instruction. If the first contact developed into a consecutive series between the two persons these were not coded as additional contacts but the overall duration of the conversation or activity was recorded. Abbreviations were used for coding action into 'speaks to', 'signals to', 'walks towards, away from' and so on. The observation procedure required the observer to follow a set of rules in categorizing his observations. The degree of observer inference in placing an action or classifying a contact was therefore kept to a minimum.

Chapter 12

Supervision: Before and After the Changeover

In this chapter we will describe the role of supervisors both in the old and new technologies, in order to assess the effect of the innovation on the supervisory pattern itself as well as drawing conclusions about the influence of supervision on the subordinate group in both situations. The requirements and practice of supervision have been the subject of much industrial research, and there is a large field of literature on this subject which permits some generalization about the characteristics of activities and their allocation and distribution.[1] While there is diversity in the nature of the supervisor's job as between industries, we can make certain statements: as, for example, that the supervisor's duties involve him in checking on the technical system to specifications laid down by the production programme and that he is accountable for this to superiors in line or specialist management. These duties can be analysed into sub-categories of control functions, general and specific, with some differences in content or emphasis depending on the nature of the technical and machine environment. His accountability for fulfilling these tasks to the limits set by the overall production strategy of the firm brings him into a communication routine with superiors, and also possibly with other collateral branches of management who are technically competent in relation to the various machine-system functions which the supervisor is required to check. Moreover, the technical expertise of supervisors is demonstrated in the special experience or formal training needed by the applicants for supervisory jobs, although in degree this will depend on the complexity of the technology.

The work performance of operatives is included in this area of check procedures, which range from routine checks on quality to detailed direction or guidance of subordinates' work tasks. The interpersonal contacts required in this class of activity will vary in frequency and, besides this quantitative aspect, it is likely that a qualitative variation in the nature of the contacts would make possible further classification along a continuum from consultative to directive or from egalitarian to autocratic styles of behaviour. The research literature has suggested again that the stage of technological development will influence this variation. Some situations may see the supervisor-worker communication develop as a form of social exchange, certain rewards being provided in return for sustained effort or extra care for quality, or in other technologies drive pressure may be directed towards employees by the supervisor.[2] Highly developed technologies may provide situational checks which make frequent rates of interpersonal communication unnecessary. The supervisor may only have to regulate the material environment of work, leaving the job activities largely in the control of employees. These are some of the suggestions which can be derived from recent research.

In the interviews before the installation of the new plant, the free-answer material gave us the first indications of how the supervisory role appeared in the perceptual field of the worker. Following this was a question on frequency of contacts with supervisors and a further direct question asking for an evaluation of the supervisory methods. After the changeover we repeated this procedure of the 'free' interview and again asked the same questions, this time with an additional request for opinions on any differences in supervision in the new factory. Lastly, observation of supervisors' behaviour in relation to their subordinates was used, before and after, as a check against the attitudes and expectations revealed in the interviews. Because the research investigation was essentially an exploration of employees' attitudes and behaviour, we shall examine supervision *from underneath,* for its effects on subordinates—rather than dealing with the supervisory peer group or with their communication upwards into the management organization. But the attitudes expressed towards supervisors almost inevitably lead on to aspects and problems of management leadership and policy, and in the next chapter these managerial influences will be described as they are seen by the men.

Before the changeover

Supervisory practices were mentioned only twice in the replies to the general question about the job at the beginning of the interview. Where attitudes towards a stimulus object are infrequently expressed, it might suggest the limited relevance or importance of the object, and in this case the effect could derive from limited contact. So here we have either of these possibilities: supervision may not be considered important in the work situation or it may not intervene in the field of action of the employees to the extent that strong attitudes, positive or negative, are developed.

Both interviews and observation revealed that contacts between the supervisors and the employees were confined to the senior member (rollerman) of the production crews. This was clearly shown by the interview replies to the question *How often are you in touch with the foreman?* and to the further question on *Who is your boss?* Operatives in the middle and junior grades had little or no contact with supervisors and thought such contacts were unnecessary: 'the foremen don't bother us and we don't bother them'. The locus of authority in relation to their task programme and to the actual operations was no further away than their own work teams. The latter question invariably drew replies explaining that 'the rollerman is the one that has the say in the mill' or 'the rollerman has got the responsibility, whatever he says we've got to obey him'. The relative autonomy of the work team through the rollermens' decisions on output, work pace, and even on starting and finishing times, created a communication channel from the supervisor to the work crew via the rollerman grade, decreasing the likelihood of direct contact between supervisors and the majority of mill grades.

An additional question requiring evaluations by the interviewees of supervisors' behaviour towards them produced uniformly favourable responses, many of which also drew attention to the very few occasions for contact between supervisor and subordinate. These replies showed that freedom from close supervision was perceived as an advantage of the production jobs: 'there are worse jobs than this. There's nobody on top of you for this and that. Here I know what to do, how much to do and nobody bothers me'.

Thus both the question on contact and subsequent observation of the crews pointed to a characteristic pattern of communication where contacts occurred almost entirely on the initiative of the rollermen who could 'call in' the supervisor, usually because of some technical problem. The rollermen interviewed described this practice in a way which revealed certitude in their expectations of supervisors and consensus between the latter and their subordinates on the definition of reciprocal duties:

'It's like this. If I think something is wrong in the mill I call him in but as for producing, it's left to me and the rest of us'.
'We don't see him at all unless we want something. He doesn't interfere unless we want something special'.

The supervisors' duties

From the observational study of work behaviour which followed the interviews it was confirmed that the supervisors' responsibilities concerned any technical faults or problems which might arise during the shift, while at the start of each shift they administered the manning arrangements, filling any temporary vacancy in the production crews. These responsibilities were shared between the senior supervisor and his two assistants. The former took charge of preparation and adjustment of the machines before the start of each week and he made daily checks on quality standards both in the product and in the raw material stocks. The allocation of crews to machines and the implementation of promotional moves ratified by the union branch were the duties of the assistant supervisors. Although operatives described contacts as infrequent, nevertheless if an operative wanted to settle any personal query (and these tended to be about pay, shift-changes, tools or equipment) then access to the supervisors' office was direct and immediate for all grades of worker.

In general the supervisors' relationship with subordinates was similar to what would be, in another context, that of a technical specialist whose services are available to production staff. The established routines of the team operations, the area of decision of the senior operatives and payment by piece-rate were the means of directing and regulating the production operations so that the supervisor provided mainly auxiliary and service functions. These factors left the supervisor in what could be considered as only an indirect relationship to the ongoing production operations and supplied the controls over task performance, which made unnecessary the use by supervisors of interpersonal communication to influence operative input or effort.

Supervisor-worker relationships

The interviews led to the conclusion that supervisors were expected not to intervene in the sequence of team operations, and we have seen how this was a valued feature of the job. As one rollerman pointed out, 'the good thing about it is that you are your own boss, nobody bothers you unless you need them', a furnaceman expressed his satisfaction in being able to 'come in and get on with the job and there's no one to boss you like in other places with the boss on top of you'. The supervisors themselves were quite explicit about these expectations, which they shared, and when they were interviewed they talked about the pattern of supervision in the same way as their subordinates. Thus behaviour which appeared to be a result of a given technological environment was also the expression of shared norms and expectations. The supervisors saw the need for approach only to the senior grade in each crew and they emphasized the value placed on autonomy in the work group, as in this supervisor's opinion that: 'Men don't like to be watched, they like to be trusted'. It was of importance here that the supervisors had spent most of their working lives in the production crews, with whom they also shared a common neighbourhood background. The production workers were observed on numerous occasions entering the supervisors' office and addressing them in first name terms without formal acknowledgement of rank. These examples of a large number of such interview replies illustrate the effects of a shared social and work experience:

'They (the supervisors) are more like mates, because they have all been through the mill. We understand one another, we know what we can do and as they have been through it themselves there's no need to bother us then'.

'Ed. was working with me . . . Sam was with me in the cold rolls. You couldn't reason years ago like you can reason with them now. I've come on with them all right, every one of them. Ed. has been right through it. They are good foremen, they are not "do this, do that". They ask you "will you do this?".'

As the evidence has shown, supervision was egalitarian in its forms and procedure. The supervisors' duties consisted mainly of service functions with initiative for these being taken by the operatives, given the autonomy of the crew. What effect would the technological innovation be likely to have on supervisory practices? One possibility of difference would derive from the complexity of the new machine system. In the old plant promotion to supervisor was from the production crews since no special technical competence was required more than that possessed by the senior operative grades. The criteria for the selection of supervisory staff would be one likely and major difference; there was also the possibility that this would lead to modifications in the supervisory roles and the question now has to be answered of how far these were acceptable or preferred by subordinates who possessed the experience and attitudes which have been described.

After the changeover

The attitude-mix which was shown on the tabulation of job satisfaction replies (see Chapter 9, Figure 12) once more does not indicate even minimal saliency of supervision in either the satisfiers or dissatisfiers. In answer to the retrospective question, when the interviewees were asked to recall the advantages or disadvantages of their previous jobs, fifteen per cent. mentioned the absence of close supervision as part of the general freedom in work which they enjoyed. That supervision emerged rather more clearly as a satisfier in the recall situation may be because it was no longer present as a positive characteristic of the present job.

In the old jobs, disciplinary limits were set by the task routines and by decisions taken in the production teams. How far was the new situation similar to this? Again, the information collected in the interviews had to be sought by a direct question; there was little spontaneous comment. Direct questions were asked on supervisory contacts, how these compared in frequency with the previous situation and, on supervisors' behaviour, how did they get on with the men and were things better or worse or just the same as before?

We will examine first the organizational changes which form the general setting for the interview responses. Especially important here are the technical requirements of supervisors, and any other discernible organizational changes in supervisory role. Afterwards, we can carry out an analysis of the interview material for evidence of the meaning of supervisory changes for the subordinate group. Lastly, differences between the two situations in terms of communication with subordinates will be examined from the interview analysis and from the observational record which was made.

Supervisory changes

The overall number of supervisory staff, methods of recruitment and the training required of supervisors were easily established differences. The ratios of supervisors to subordinates had increased from 3:50 to 4:30, almost doubling the number of supervisors for each employee. The increase in line supervision alone did not, however, represent the total numbers of supervisory staff since there were also the maintenance supervisors and technical staff who, while not in a position of direct control over employees, nevertheless exercised authority through their knowledge of the machine system. The transferees were used to a single line of command and, at the beginning, found the increase of supervisory staff confusing. Several employees expressed similar complaints to that of the crane driver who narrated how 'orders were given to me by people I had not seen before, I didn't know who they were or which position they had'.

In their general background of experience as well as in special training the supervisory positions had changed to the extent that senior supervisors of the production department were university graduates and others among the assistant supervisors had technical qualifications. Because their appointment might depend on these and not on their experience of the work operations, they were much younger than supervisors in the old plant. The old-type foreman came into his position after a long period of service as a rollerman or furnaceman and he was usually in his fifties. Now the supervisor was appointed only a short time after attaining the formal qualifications. The shared experience of operatives and supervisors had therefore disappeared and there were other points of separation: because of the much wider geographical scatter of the recruitment area, in only a very few cases was there any mutual connection between supervisor and subordinate through residence in the same community.

The separation which resulted from these differences might have been expected to emerge in the mens' attitudes. On the other hand, in a technology with the pre-scheduled and fixed machine programmes of the present system we would also expect relationships between supervisor and subordinate to be relatively easy or certainly without the pressures or drive on subordinates that have been demonstrated in assembly-line or short-cycle repetitive work.

However, the technological innovation had also introduced a zone of very great modification: the operations were on a much vaster and more complex scale, and a breakdown at any point was an emergency for the whole process. These were the sources of new definitions of the roles and skills of supervisors. What were the consequences for the incidence and content of communication between supervisors and the labour force? What were the nature and meaning of the changes for their subordinates? The answers to these questions are found in the interview responses and the observational study of communication between production workers and supervisors. By taking the two types of evidence together, the supervisory styles of the new factory can be assessed with some interesting results.

More frequent communication

The first question asked on supervision required a comparison of the frequency of contacts between supervisors and the interviewees.

Table 25. Are your contacts with supervisors more, or less, or about the same as before? ($N = 51$)

Less		*No difference*		*More*	
No.	*Per cent.*	*No.*	*Per cent.*	*No.*	*Per cent.*
10	20	13	25	28	55

Twenty per cent. thought they had less contact with supervisors: these tended to be the transferees who had moved from senior grades in the old jobs down to ancillary or labouring jobs, and it was this change in their own position in the operative hierarchy which was more important in producing the decrease in contacts rather than the supervisory patterns *per se*. Twenty-five per cent. saw no difference one way or the other, perceiving generally that they had as little contact now as in the previous situation. But for fifty-five per cent., supervisory contacts were now more frequent than before and this category included all those in production line jobs. For example, in the following replies the interviewees explain that 'it's different now, we don't have one we have many', or 'there are more of them, in the old works there was one supervisor', or 'there was only one supervisor before, now I have five'. For six of the interviewees the mere proximity of supervisors in greater numbers than they were used to before was sufficient to give the impression of more 'contact'; these men reported 'seeing more of them, they are around all the time'. The more numerous supervisory staff spent more time near the production lines, and proximity in itself would increase the likelihood of verbal communication.

Observational recording: the pattern of communication

From the observational recording of four production teams, each for twenty five hours, we found a characteristic pattern in supervisory communication, which was similar to that of the old plant in its tendency to be confined to the top two or three operative positions. This was not the absolute restriction of communication to the senior operative as was the case in the former crews and there was now a wider distribution of contacts over all grades. Nevertheless, there tended to be a concentration on the upper grades: for example, sixty per cent. of the contacts initiated by the supervisors to the team of nine members were directed to one or other of the two senior operatives (one hundred and seventy-five out of two hundred and ninety-two contacts). The similarity ends with this.

What constitutes the major difference in the new plant is that the supervisor *initiated* communications much more. Instead of waiting to be alerted or 'called in' by the senior operative, the supervisor now initiated contacts in more than half of those observed, and in certain teams in a heavier majority of instances.

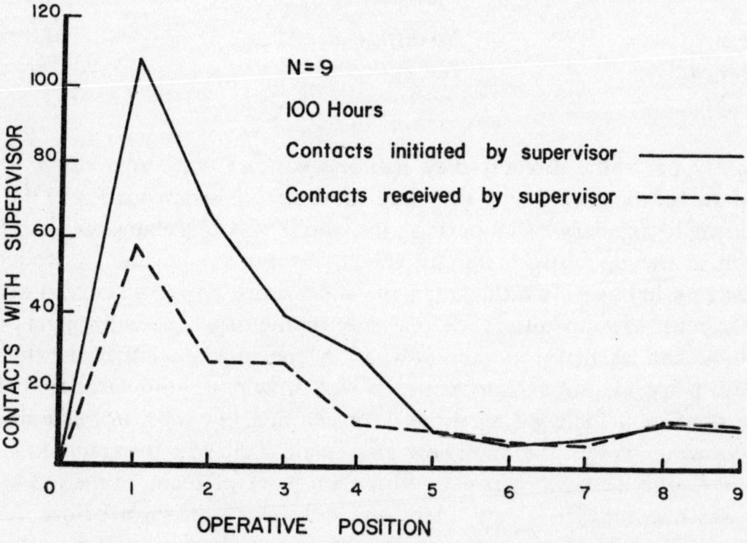

Figure 15. Supervisory Communication to Production Team

Thus, the analysis demonstrates a communication pattern which tended to be directive, that is, initiated by the supervisor, and in many cases without requiring a reply in return from the operative. Figure 15 illustrates the preponderance of contacts initiated over those received by supervisors in the four teams whose communication rates between operatives were illustrated in the previous chapter.

Qualitative differences

As well as the overall increase in frequency there were evidently marked qualitative differences in the nature of supervisory communication. The men described this situation where the supervisors . . . 'behaved differently, now they only talk to you to give you a command to do something '. The larger technical responsibility of the supervisor and his more frequent checks on the operation of the machine system were indicated; the nature of the authority vested in the role had changed and augmented.

In particular the locus of authority was now removed from the work crew because of the supervisor's power of decision and check. The operative routines in the old plant were more well-defined and, as a line attendant recalled, 'then the foremen weren't responsible for production, we were'. Others made a similar comparison, drawing attention to the greater degree of operatives' control in the less complex technology:

> 'Then you knew your work and got on with it. It was more like routine in the old works, here the foreman gets the first information on it'.
> 'Foremen in the old works didn't have much to do with you. It was repetitive work, you only saw a foreman if there was a change of order or there was a breakdown, but in this automated plant foremen have to be around all the time . . . the foreman is liable to take it out on you if there is a breakdown, not like before, then it was obvious what was wrong'.

The fact that the discretionary area which used to belong to the operative was reduced, and the authority of supervisors increased, might suggest the prediction that there would be an awareness on the part of the operatives of a stricter application of general rules, and we will examine this possibility later.

Let us return to the data on the communication pattern. Thirty-four per cent. of the interviewees in answering the question on contacts with supervisors said that these occurred only within the procedures of check on the machine system. This was in contrast to the communication content in the old plant. The interviewees drew attention to the present situation where 'the foreman talks mostly about the work, nothing much besides that. Not like the social chats we used to have before with the foreman'. Supervisors used to be 'more sociable' and 'would talk about things other than work'. An interviewee compared the present and previous relationships, explaining the effect of the new procedures:

> In the old plant you were closely connected with them and you could have a chat about anything, but now you can't approach the supervisor the same. You have to have a concrete thing to talk to them about, there's no amicable talk between you and them'.

It became clear that communication had changed to the extent that it was confined to work topics, and this presented a departure from the previous relationship where the supervisors were 'more familiar'. The role of the supervisors now appeared to incorporate greater distance and formality—partly because of their new technical authority. We can examine the interview responses for further light on the differences in relationships. Certainly, any changes which the men encountered, for example in the increase of communication downward, did not derive from *personal* characteristics of individual supervisors. Answers to the question *Do you get along as well with the supervisors here as in the old plant, or better or worse?* showed the following distribution:

Table 26. Relationships with Supervisors (*N* = 47)

No difference		*Better*		*Worse*	
No.	*Per cent.*	*No.*	*Per cent.*	*No.*	*Per cent.*
29	61	5	11	13	28

Eleven per cent. thought that relationships with the supervisors in the new plant were better than before and sixty-one per cent. could find no difference in this respect: 'No-one is unpleasant here, they're pretty good'. Twenty-eight per cent. thought the supervisors got along worse with the men although the negative opinions were expressed, not in personal terms or referring to individuals but against the supervisory procedures in general and particularly the impersonal quality of these. 'They treat us well but there is not the same familiarity' was typical of replies in this category.

Observational analysis corroborated the large quantity of interview material in establishing the nature of the changes in supervisor behaviour, which was now more directive. Decision-making was not shared with the operatives as it was in the old production crews. The greater technical expertise required in supervisors prevented this happening. A measure of applied-science training now tended to be a condition for selection to supervisory posts whereas previously the supervisors were appointed from among the experienced workers. One interviewee expressed this difference briefly and adequately: 'Now, other people, the foremen, do the thinking for us'.

Such a development would tend to assist the separation between supervisor and operatives and formalization of role, and to this extent the differences in the relationship had derived from the *mise-en-scène* of the continuous flow process. But it was unlikely that they were the results of this alone, and a question immediately presents itself on how far the supervisory roles as part of the management organization of the firm were themselves structured or formed by managerial policy.

Managerial policy and supervisory behaviour

There were some signs that the men were aware of a general policy, implemented by the management: for example, in the references to the arrangement of shifts which placed supervisors on different rotas to the work team, so that no supervisor remained with any one team on consecutive shifts. This was seen as part of an effort to maintain distance 'for the foremen not to get too friendly with the men', and it was confirmed to the researcher later by management members who pointed out the importance of the rota system in maintaining the supervisors' detachment from individuals in the teams. They considered it a necessary condition for fair supervisory practice.

The value of this sort of policy is given some plausibility in studies on leadership behaviour which have suggested the hypothesis that more successful results in task behaviour are achieved when a certain social distance is maintained between leader and subordinates. On the other hand, it could be argued that this might diminish opportunities for optimizing leadership influence in communicating rewards to subordinates, whether these are in the form of encouragement or guidance, or information on progress.[3]

The end-results for operatives

It was concluded from the evidence that a formalization of role in the new factory had introduced distance in relation to operatives at the expense of a support function from supervisors. Furthermore, it was strongly suggested that such a function might even be especially needed after majot innovation, given the presence of other influences which were shown in the last chapter as working against the formation of well-defined interpersonal roles among the operatives themselves.

A situational effect which made this kind of support function less possible was the difficulty encountered by management and supervisors in locating sources of stoppages or failure in the machine line. It emerged in forms of pressure on the supervisors similar to those mentioned by Faunce in his interviews with employees of a newly mechanized automobile plant.[4] The steel workers described the occasions when these stoppages occurred, and supervisors sought to allocate blame onto members of the production crew:

'You've got to be on your job all the time, because if anything happens they've got to be prepared to blame someone else rather than take the can back themselves. Here they are ready to pass the blame. They are more afraid of their jobs here than anywhere else, if they made mistakes all the time, they might lose their jobs'.

As we have seen, the pattern which had developed was one of continual proximity of supervisors to the men, but with the exception of men in maintenance gangs, whose attitudes were more favourable, proximity did not lead to ease of approach or to relaxed styles of supervisory behaviour. Moreover, the finding that distance was increased ran counter to those predictions of the *typical* supervisory pattern in automated plants.

Indifference to supervisors: A summary

Many of the men were indifferent to supervisory practice both in the old and new plants as far as it affected their liking or disliking the job. Supervision did not attract opinions in the strength, negative or positive, which was evoked by other factors. But the retrospective questions (see Figure 12 L.2.) drew out the men's preferences for the former supervisory practices particularly as they were associated with freedom in the task routines: the men were not 'bothered' and there was 'no interference' in the way they carried out their jobs. In that situation, contact with supervisors might occur at the beginning and end of a shift but not during the shift itself, and many of the operatives below the grade of rollerman had little or no communication with supervisors.

To sum up: in the new factory the men perceived that pressures on supervisors were greater because of the duties of the latter in machine scheduling to secure production at maximum capacity. The supervisors now communicated more with operatives but there was no corresponding flow of communication upward, with fewer instances of communication actually started by the operative. In some of the production teams, supervisors communicated to operatives twice as much as the latter initiated communication with them, and in all the teams communication downward predominated. This was the greatest difference between the two supervisory situations.

While there was no intensity of criticism, there were some unfavourable references to an increase in formal separation between operative and supervisory levels: the supervisors were less approachable and contacts with subordinates were limited to technical matters; in effect, there was altogether more distance in the hierarchy.

One source of differentiation was the technical education of the new supervisory staff and their dissimilar background in work experience. But it was also clear that supervisory procedures were standardized in the company and appeared the same to employees in different departments, so that it could be concluded that managerial policy was directly involved. The latter was expressed, for example, in the back-rota system, introduced by management with the purpose of preventing supervisors from over-identifying with (and possibly favouring) any one production team: a system seen by interviewees as a method of 'keeping supervisors from fraternising with the men'.

Some questions

There are several questions raised by the data collected on supervision. In the first place, the bare facts of the case: in the automated process the supervisory pattern, for the various reasons outlined, has become more formal, distant or directive. It is not necessary to pass a value-judgement on this development, but the findings are worth emphasizing because they do not support the predictions that supervisors and operatives become parts of a close team relationship in the new technologies. From the men's point of view, which is essentially what is being reported in this book, they were in fact further away than previously.

In conclusion: at this intermediate stage towards full automation (which is where most automated processes are) there was no evidence of the supervisory relationship with employees becoming closer through shared responsibility. The interviewees perceived a managerial attempt to maintain formal procedures and distance between supervisor and subordinate. What was the pay-off for supervisory distance? Although preventing supervisors from being influenced by their subordinates, it might also hinder the supervisors' influence over them and minimize a reciprocal information flow. Might the operating unit be considered as being more effective if operatives' expectations and supervisory behaviour are more closely geared together? Could management have given more recognition to the controls which might have been, in any case, inherent to the new technology allowing a less directive pattern of supervisory practice?

REFERENCES

1. Supervisory behaviour is discussed extensively in Bernard M. Bass, *Leadership, Psychology and Organisational Behaviour,* Harper & Row, New York, Evanston, London, 1965.
2. The relationship between technology and the organizational tasks of supervisory staff is considered in Joan Woodward, *Industrial Organisation: Theory and Practice,* Oxford University Press, 1965.
3. A highly relevant discussion of ratios of communication between supervisor and subordinate is found in Mason Haire (Ed.) *Modern Organisation Theory,* John Wiley, New York, 1965, pp. 162-164.
4. William A. Faunce, 'Automation and the Automobile Worker', *American Sociological Review,* 23, No. 4 (1958), p. 401.

Chapter 13

Management: Before and after the Changeover

Observation over the period of eighteen months inside the old factory revealed a characteristic style of management behaviour towards the labour force. This evidence emphatically excluded the notion that there was frequent contact between manager and worker, and it demolished the supposition of a smooth conflict-free relationship in the traditional-type small factory.

Despite the organizational simplicity of the company and the physical proximity of the administrative offices to the workshops, senior managers had no routine of visits to the factory. One worker expressed the general opinion when he said: 'I've never seen the manager here in the works in three and a half years'. This pattern of management could be expected where the technical and output controls were located in the operating crew, and the absence of management intervention produced the effect which the men called 'the no-discipline'.

However, the adoption by the general manager and other senior managers of behaviour which produced an attenuation of all workshop contacts with employees led to some further interesting consequences in the attitudes of the labour force. Worker representatives who reported back on meetings with management generally supplied the personal details which composed the current stereotypes of management and stimulated a characteristic tone of industrial relations in this factory.

Stereotypes about management

This situation illustrated very clearly the influence of secondhand opinion on the formation of attitudes. The traditions of the earlier history of the firm, in particular, seemed to reflect a dominance-submissive type of struggle between unions and management: deriving from a repeated series of attempts at dominance by one party or the other. At the time of the investigation the militant local union branch had gained some advantages for its members and shortages of skilled labour worked in their favour. Illustrations of past managerial policies were narrated during the interviews with a description of incidents which, under the usual conditions of social learning, would have been passed on to the new recruits. The following are typical examples of how the men recalled a former and extremely severe disciplinary code:

'There were hard times here when they had the whip. I've seen a grown man crying after breaking a roll. I remember men terrified of going to see the manager. I remember one saying "I knew something was going to happen, a picture fell in the house last night", and things like that. When my father came in after a roll had broken it was as if there was a corpse in the house. But we have come a long way to better ourselves here. We've had a real strong branch and a real active committee'.

In one of the branch meetings which the writer attended, complaints by an official about the manner of the general manager's rejection of a wage claim led to criticism of the management reaching a crescendo of strongly personal insults and abuse. Eventually only the branch secretary's strenuous efforts enabled translation of a number of abusive personal adjectives into moderate forms suitable for recording in the minute book.

Both content analysis of branch records and attendance at the meetings pointed to the union officials representing and leading the opposition of the employees against management policies, but at the same time restraining the more extreme of their members. By conveying to members information given by management on market conditions and prospects and in fashioning the more extreme antagonisms of members into workable resolutions, the officials were engaging in a dual role.[1] Very often their position in community associations where they were office-holders together with senior managers assisted this duality. Just as shared community interests and leisure group memberships lessened the differentiating effect of the formal structure inside the factory, the ties of kinship and neighbourhood may similarly have minimized the consequences of management-union disagreements. Within the conventions of the dispute procedures conflict was present and was verbalized to an acute degree, but antagonism and opposition to management actions expressed in union meetings was rarely brought to the point of a strike: in the previous twenty years this had happened only once, on the occasion of a dispute general to the industry.

Management non-intervention

In the old technology there was no management drive on employees, this would have been impracticable because of piece-rate payment and the agreed output levels. No system of check was possible on the skills of judgement and discretion; given the importance for successful production of the direct decisions by employees, management was not required to intervene and immediate supervision of production was largely in the hands of senior workers.

The management left the work crews to run themselves; a further factor in management-worker behaviour was the control by the union over the manning arrangements of the crews. The branch administered the filling of vacancies and promotions, setting work performance and even disciplinary standards. As was shown in Chapter 3, the branch had taken over many of the internal management functions. In this situation the management skills and attention were directed away from the internal running of the enterprise. As in Simon's illustrations of the behaviour of the head of the enterprise, the 'external focus' of top management was directed to selling, to market conditions and to the obtaining of raw materials and orders.[2]

Access to management

For the employee even visual contact with departmental managers was infrequent; with supervisors communication occurred as or when it was functionally required because of technical problems–and on the initiative of the subordinate. Yet in another sense management members were not remote. An employee, at his individual request, could obtain access to the senior managers at any time, although the frequency with which this might happen was low because the occasion for access was some exceptional personal problem. But the reactions of those few in the sample who had taken this line of access to the head of the firm indicated that they had been favourably impressed and that the impression continued for long afterwards.

At the same time, there was some indication that the prolonged absences of top management from the mills did not really fit employees' definitions of the general manager's role. It was possible to ask only eleven of the production workers the direct question on whether more frequent top management appearances were thought necessary or desirable, and of these seven replied in the affirmative. Their opinions showed a tendency for operatives to give approval to certain forms of behaviour on the part of the general manager; their expectations were not exacting and clearly did not require more than his occasional presence in the workshops: that he should be seen to 'take an interest in the work and in the men' was the extent of the statements that were made.

This minimal degree of orientation towards the top management hardly reflects a condition of dependence of the type which has been suggested as inherent in behaviour between managers and workers.[3] The notion that operatives are heavily dependent on their superiors was not relevant to the steel workers' situation of multiple loyalties and group memberships in the firm, such a notion would miss out the functions both of the union and the peer group. Because of the control over work activities which was possible from these two sources, the steel workers were anything but dependent, with the connotation of helplessness that the term might imply.

However, in the current conditions of rumour there was an increased emphasis on the importance of the general manager's role in relation to the common goal of continuing production in the obsolescent industry, and he had power to reassure on future policies and to resolve uncertainties. Seeking information on their immediate future was a continual preoccupation of workers in the old plant. Thus, the employees appeared to engage in scanning for signals from the top of the authority structure. The worker's projections onto top management of the power and will to solve problems of inefficiency has been described by Lupton,[4] in this case where top management was thought to have the power to 'put right' a situation. Lupton's study gave an insight into the very real significance with which the top position may be invested. For the steel workers the information content of any communication from the headship of the firm was of particular importance in the prevailing market and technological conditions, although no direct workshop control was required of management and the men valued their freedom from close supervision.

After the changeover

We have mentioned the organizational changes in the new plant, the increase in technical complexity and in specialist staff. In the looser and technically more simple organization of the old plant, both managers and supervisors could afford to be disengaged at the operational level. Now the development of more refined techniques has meant the incorporation of specialized plant-operating skills into the management and supervisory structure and a more direct responsibility on their part for the machine operations. It is not within our brief to describe these managerial changes in themselves but to examine them from the viewpoint of those in the subordinate levels of the organization. Since the introduction of the new technology, what were the alterations in managerial roles which were perceived by the workers?

The structural changes in a more highly specialized staff, the higher ratios of supervisors and managers to men: these might lead to observable differences in the management-worker relationship. Content and frequency of communication between management and the labour force would be one obvious behavioural variable which we could examine as a reflector of these structural differences. Other influences on the relationship might not derive directly from structure in itself but from what the interviewees perceived as a different management ethos.

The interviewees were asked a general question, *How much do you see of management?* and this was followed in the majority of interviews by a question asking for comparisons. These two questions, with follow-up for more detail on specific points, gave information on frequency and quality of management interaction with the work force and on how management activities were perceived by employees. Inevitably, the comparison question drew out evaluative comment on both situations. This indirect approach sufficed to give an indicator both of the strength of feeling attached to the relationship with management and of the characteristics or roles which elicited positive or negative responses from the men.

Management behaviour in the new factory

That the concept of efficiency involves individual and social factors as well as economic considerations is well established as a basic assumption in industrial studies. There is also a body of opinion and evidence in the literature which converges on the statement of managerial responsibility for social as well as technical coordination. Following on from this, it has been argued that company policies should take into account a wide range of rewards which might be obtained or expected by the individual worker.[5] What can be said about the success of the managerial role in the social organization of the new factory?

The greater number of managerial staff and their increased technical duties had the effect that managers were frequently present around the production lines and were often likely to be in the proximity of operatives. Top management made routine tours of inspection and the men described this *visibility* in terms which emphasized by contrast their absence from the area of operations in the old works. There were several comparisons like the following:

'In the old plant you would never see the manager. You'd practically never know who he was, here you have them around every day'.
'The manager didn't come out from his office before, and if he did it was because of a row'.

The men were here referring to the head of the firm in the old plant and to his preference for 'ruling things from a chair', this pattern of behaviour sometimes being interrupted on the occasion of a supervisor-worker dispute when he would 'rush out into the works like a bull'.

Before the changeover the apex of the management structure was this general manager position; below it were the managers responsible for a number of locationally, technically and administratively separate departments. In the interviews the men would refer almost always to one of these two positions (i.e. general or departmental head) and only to these. Now, however, while they mentioned a variety of line or technical positions, the replies also showed that there was still an orientation towards the head of the firm, and the new general manager was frequently discussed in a more positive way than the holder of the equivalent position in the old factory.

The following conclusions can be drawn from these opinions. Specifically, what may be expected by employees of the occupant of the headship position in a company is no more than a favourable presence. It is important to the worker that the head of the authority structure transmits sufficient cues to indicate that those with power over the work environment are basically well-disposed towards him. Visibility may be all that is required of top management, together with fairly minimal expressions of interest. In the case of the transferred workers, this class of expectation was easily satisfied, particularly in contrast with the previous experience of spasmodic and less predictable top management behaviour. One worker expressed the difference between the former and present general managers with the following illustration:

'This one will say "good morning" whereas the other wouldn't say anything if he fell over you'.

Absence of community ties

Aside from the top position and in relation to the larger number of managers in the intermediate levels down to supervisor, certain negative effects appeared. As between manager and worker, communication upward in terms of direct access was more difficult than before. Relationships were perceived as more impersonal and by the fact that most of the managerial staff did not share the immediate neighbourhood background, there was a loss of the underlying informalities. The effect of the local community ties had been to produce acquaintance with managers outside of their formal roles in the factory and to blur authority lines:

'You were all one community from the manager down. Everybody knew everybody . . . you were more familiar than to these. You knew them. In the old works you could stop the departmental manager on the floor. We knew him and he knew everyone, not like here, here you're a number, nothing else'.

Formal procedures

Besides the absence of this broad basis of informal relationships with managers or supervisors there was also the greater structural separation from management which the men described. The communication pathways were not so direct and formal procedures could not be by-passed through *personal access* to departmental or senior managers. Interviewees who drew attention to the lesser likelihood of any personal contact, gave as a reason the use by the management of intervening supervisory levels:

'They don't want to bother, they get their understudies to do that. You don't talk to the

staff like you used to, they don't expect it, they just pass through'.
'We see them more often but they get to us through the foreman'.

Thus, towards middle management there was a perception of distance. The departmental managers were seen around in the plant but there were no forms of communication which could be compared with the more direct contact possibilities in the old plant.

The men experienced difficulty of approach to managers, and they did not perceive the procedures for union representation of the individual case or the frequent communications from supervisors as substituting for the formal separation from management. Communication upward was so organized or guided that there could be no possibility of an operative's work grievance having any rapid resolution; the previous situation again gave the frame of reference to the men's reaction to existing procedures:

'You are further away than before. Before you could meet the manager when you wanted to, not now. The company makes it that way, you are not able to meet them unless you take it through the union. For the company it is a good thing for you not to get into the engineer's office. For the workers it is bad not to be able to meet management more often and get things settled at once'.
'Before you could give your opinion on anything and you could always approach the foreman or go right to the top. Here it goes up and up (with gesture illustrating a number of ascending steps) until it's squashed!'.

Perceptions of a management ethos

There was a strong trend manifest in the interviews to see, more than the sheer structural changes, a management ethos which was expressed in a particular behaviour style. This was noticeable in the men's description of the 'regimentation' which, it seemed to them, characterized their relationship with management. For the employees it was now more sharply clear that managers had sentiments and interests which were not shared with the work force. 'These people here don't bother with us at all. I suppose it's because they feel differently to us'—this awareness in one production worker of differentiation and social distance takes us further than task-required separation or the differences deriving from the specialized technical education of the newer staff. It expressed the imputation of a set of values which were thought to be implemented in the behaviour of managers.

A few of the men compared the prevailing management culture with a military hierarchy of ranks. A union official, discussing management personalities, described them in a generally favourable way but added, 'No, it isn't that they don't want to be friendly, it's that they can't, they are too ingrained in army routine'. The production workers particularly appeared to encounter 'regimentation' in the established management procedures, and were inclined to talk about the managerial influence on supervisors as producing greater distance between these lower levels of staff and the employees.

The comparison was made between day and night shifts in this respect. During the day, formalization of supervisory roles was held to be more evident for the reason that the various management levels were in full complement at this time:

'Some of the foremen are not so friendly as in the old works. I suppose it's the management here, they are more regimental . . . especially on the day shift when the big men are walking around. In the evening the foremen themselves are happier, they are happier themselves from 6 in the evening till 9 in the morning'.

Conclusion

To summarize, the top management emerged as carrying out a role which was more acceptable to the men than was its counterpart before the change. But for the intervening management levels there were complaints of more difficult access to managers. Distance certainly existed in the old plant, taking frequency of contact as the criterion, but the work group was virtually autonomous technically and the remoteness of management was part of this autonomy. Formerly access to top management by individuals had been available, whereas the increase in managerial staff was now accompanied by procedural rules which discouraged direct access. The loss of the operatives' manipulative skills and the limitations in their knowledge about the technical system was also associated with this separation between levels in the organizational hierarchy.

The middle management behaviour pattern was very different from the old factory insofar as managers were now continually present in proximity to the operatives; they were *seen* frequently but there was little communication between them and employees. Among the transferees this behaviour was not taken as a legitimate organizational requirement deriving from the rationale of the technology. A tendency was shown to see the new managerial roles as originating in a deliberate policy to maintain formality and distance. What might be an organizational effect was therefore perceived as a management strategy.

This was a fundamental change because distance between employees and managers in the old factory developed through the absence of the latter from the workshop during long periods. Now, middle management were required by their technical role to be in the plant continually, but although this was so there had developed an excessive formalization of role. It no doubt accounted for the curiously faceless impression of management which emerged from the interviews. Taken on its own, this was not among the most salient of the many topics which employees raised in the interviews, but it did connect up with wider social issues in the factory.

Summary

In the old plant the union had gained control of internal manning and administering of the production crews. The tone of industrial relations was militant following an earlier period of more severe managerial discipline. Direct access to managers was possible by individual workers although the notion of a dependency relationship between worker and manager was not upheld by the evidence, in view of the solidarity of relationships among the membership of the local union and in the production crews. However, the general manager's role before the changeover was given particular importance by employees because of his perceived power to resolve employment uncertainties.

The increased technical complexity of the new plant required a more numerous and specialized managerial staff who were often present in the workshop although there was very little direct contact between them and production workers. Opinions towards the general manager were found to be more positive than towards middle levels, and this was again interpreted in the light of expectations held of the headship of the firm. Opportunities for access of the direct kind possible in the old plant did not exist. Disappearance of community cross-ties partially accounted for this but there appeared also to be a management policy of greater formalization between organizational levels.

REFERENCES

1. For an exposition of the role of trade union officials, on the one hand for their members, and on the other hand for management, see Michael Fogarty, *The Rules of Work,* Chapman, London, 1963. The 'dual function' as expounded by Fogarty consists of furthering members' interests while concurrently assisting in the productive goals of the enterprise.
2. Herbert A. Simon, *Administrative Behaviour,* 2nd edition, The Free Press, New York, 1965.
3. See Chris Argyris, 'Understanding Human Behaviour in Organisations: One Viewpoint', Chapter 5 in Mason Haire (Ed.) *Modern Organisational Theory,* John Wiley, New York, 1965, pp. 115-154. Argyris discusses the factors producing dependence relationships in a work organization.
4. Tom Lupton, *On the Shop Floor,* Pergamon Press, Oxford, 1963, pp. 193-194. In this study supervisors were criticized by workers while top management completely escaped criticisms: 'We often heard the remark "if only they knew what went on down here they would soon put a stop to it".' Lupton concludes that in this case the selectivity and direction of criticism was realistic.
5. A discussion of the 'rights, privileges and obligations between worker and organization' is given in Edgar H. Schein, *Organizational Psychology,* Prentice-Hall, Englewood Cliffs, New Jersey, 1965, pp. 10-13, 43-65.

Chapter 14

Conclusion

Retraining and advance preparation

A recurrent question during innovatory change concerns the suitability of older employees for re-training as production workers. In this study it was shown that a bar on those over forty-five years of age for production jobs led to chronic dissatisfaction among men who were near but above this age bar. Not only was this so, it was suggested after analysis of the skills and requirements of the new jobs that these men should not have been debarred on grounds of age alone. Evidence on the difficulty of jobs in the automated plant did not hold out the possibility that they required any particular set of capacities which, it could be claimed, older workers did not possess. In the steel industry the accumulation of experience in long service employees might well constitute an advantage, even where machine changes are very large. At a minimum, it would provide no interference from previous work-habits on learning the new production methods.

The possibilities for training schemes of a more general educative nature than job instruction of the conventional type were suggested by the interest which the automatic process had for the operative as well as by the job analysis. Although his tasks were not at a high difficulty level, the instrumentation of the machine line was complicated and some knowledge of technical components and conditions would have introduced more meaning into the job content. Search strategies which were shown to be important in job performance could also have been improved if opportunities had existed for technical learning.

By recognizing the necessity for production workers to have more technical knowledge, the way would be clear for several outcomes: the individual's interest in his job would increase, and an enlargement of his job role and area of action could, in turn, lead to an unburdening in supervisory and management duties towards operatives. One accompaniment of making explicit the actual change in job content might also be an alteration in the rationale for wage payment. Traditionally the latter has been based on the notion of payment for effort in metric units of time, but it would follow changes in operative role if wage payments now moved towards a similar basis as managerial and white-collar occupations (i.e. monthly or even yearly salaries), as has already happened in the oil and chemical industries.

No-one would argue against the advisability of preparation of the labour force in advance of major industrial change, although research may still be carried out on the methods used, for example in training or consultation with employees. The importance of preparation whether by advance information or by re-training schemes has been demonstrated in much of the published evidence. This seems the one indisputable case where it is possible to affirm the value of special 'worker communications programmes' which have been criticized in other contexts[1] but may have an effective application during the period of introducing employees to the details of impending changes. However, one of the well-established conclusions of the human resources research of the Tavistock Institute was that more fundamental policies were needed.[2] In terms of priorities, communication procedures will follow planning for the organization as an entirety with the emphasis on *behaviour* changes for operatives, supervisors and managers, to be estimated and specified concurrently with the engineering design.

In the literature on technological change it has occasionally been possible to encounter an inappropriate assumption–namely, that employees must adjust; and if they do not, it is said to be because of resistance to change. It is now becoming recognized that adjustment can be engineered in a different direction: superior-subordinate relationships; the content of jobs; the range of rewards; the forms that the work group may take–these may be included in design formulation. Davis has cited examples of job design alternatives which relate to the individual's control over his behaviour in the industrial plant.[3] The illustrations and findings quoted by this author confirm that the designer and the social scientist can work jointly with the aim of optimizing both the engineering and social policies of the firm. For the latter, it would be largely a problem of finding ways to arrange design in order to achieve an appropriate social result. This does not mean that we would arrive at sub-optimal technical performance. It does mean that the machine system will specifically take into account and provide for the work team as a nucleus of connected jobs.

Job incentives and the primary group

There are conflicting opinions still about the outcome of interaction between technological and social variables. It seems that in automated production the primary group may be or is expected to be either more cohesive or more isolating for the individual than at earlier stages of technology. The latter result has been observed for the steel workers and it will be discussed below with some other topics which are relevant, such as supervisory styles and the process of adjustment to new jobs.

The increased formalization in relationships between supervisor and worker which I described after the transfers to the new plant can be compared with a similar change reported by Faunce, following machine innovation in motor engine assembly.[4] In both situations innovation was associated with an increase of directive controls, from the staff hierarchy immediately above the operating team. This gives an interesting contrast to the oil and chemical industries where styles of supervision have been encountered which are relaxed and assume something of an advisory relationship towards the operative.[5] One explanation might be found in the rate of change within these different industries. Oil and chemical technologies as well as being more advanced in automatic methods have undergone no recent innovation in equipment or personnel, and this may give a clue to the differing supervisory or managerial styles. In other respects the steel and oil industries have much in common: continuous flow production, a technically sophisticated management and a labour force engaged in monitoring or maintenance. These technological common factors, however, have not constituted a sufficient influence for similar behavioural effects to be produced.

In Chapter 9 the individual's adjustment was analysed into several stages of anticipation followed by task and social learning. We saw how the men who were interviewed related their expectations realistically to the situational constraints. It was inferred that there was a process of modifying expectations to match what was actually available from the present jobs. But certain of the interviewees' values and expectations had not modified: there was a persistence of dissatisfactions with the immediate job environment and with social changes. The tedium of the continuous line, the disappearance of a structure of satisfying social relations, were the main topics of unfavourable attitudes after innovation. It is important to emphasize that these were not separate issues; for example, the isolation of individuals from reciprocal activities of work crews was a factor contributing to their state of boredom. In many instances earlier in the book we have seen that primary group behaviour can be of significance in influencing job content and work load.

How, specifically, did the effects of innovation appear in social behaviour inside the plant? Transferring the employees from one technology to another produced a break-up of established social routines, leaving the individual uncertain about courses of action which he should take and the responses from others which might follow. Interpersonal contacts in the new factory occurred with lower frequency and duration; qualitatively they provided less opportunities than formerly for task sharing or job exchanges.

It is conceivable that under the stress of innovatory displacement from jobs and established relationships, the individual became more aware of the material usefulness to him of these forms of reciprocal behaviour; but industrial employees are not a special case in respect of the importance they attach to norms and routines for reciprocal cover, and this type of behaviour has been found in contrasting occupational environments. For example, Nigel Walker illustrated how far they were important for clerical officers in government service. He described instances in particular where the officials exhibited solidarity both with subordinates and immediate colleagues, expressed in the tendency to cover up for others, to help each other out of difficulties or to defend colleagues against criticism from outsiders.[6] Similarly, examples of cover activities were observed by Isobel Menzies among nurses at a London teaching hospital. A variety of collusive norms and actions appeared which were termed the 'social defence mechanisms' of the nursing staff.[7]

It is clear that while members of groups in widely differing environments may give importance to such forms of cover activities, it is just as likely that in other occupations behaviour may be allowed which is appreciably more competitive, personally advantageous at the expense of collaborative routine. Factors which might be decisive or contributory to these differences are the career opportunities in an occupation and its promotional hierarchies (restricted but prized opportunities might be associated with competitiveness) or else the expectations which individuals derive from reference groups outside of work.[8] It will therefore be unlikely that we can make generalizations about the optimal primary group structure to suit any environment or occupation but rather an attempt can be made to consider the appropriate forms which are *relative* to actual situations. In face of innovation it would be possible to appraise existing social behaviour and then to consider alternative structures with the aim of optimizing them in the machine and organizational design.

Work group design

For the remainder of the chapter we shall discuss the practical and theoretical possibilities of giving detailed attention to social behaviour as part of forward planning. We may want to consider how such behaviour may affect the tasks and operational content of work or to investigate other effects of social engineering in the industrial firm. Can hypotheses taken from research on laboratory or natural groups be applied to design? For example, the relationships between rates of contact and the development of attitudes might be tested by experimental introduction of behavioural changes whether within a group or between subordinates and those in leadership roles.

Innovating at the level of the work group ideally would involve an appraisal of social structures already existing in order to discover which elements might be maintained, discarded or improved in relation to the alternatives possible in a new design plan. Could this approach have been applied in the steel industry on the basis of the description which the before-changeover study had provided? For example, we could examine the data to consider whether there were any behaviour elements of the old plant which might have been replaced by improved forms at the changeover. The promotional sequence gives an illustration of what might be possible. Although there was an assured promotional progression over the long term, this could have been converted to a shorter overall period. Again, the seniority procedures which were cumbersome and sometimes protracted seemed to offer possibilities of improvement.

The problems presented would have been extremely complex because of the controls exercised by the local union branch in workshop administration. To take another example, choice of a team by the individual was a major feature of production crews and it could be suggested that this characteristic of group behaviour might have been maintained. But it was implemented through the local union meetings and even if it had been given attention at the planning stage as a positive behavioural mode, management might have found it inexpedient to allow its continuation in existing forms. Would it now have been administered by management or by the unions?

The promotional ladder also had implications for union loyalties. In Chapter 8 we saw that the innovation, or more accurately the method of its introduction, decreased opportunities to move up the grade hierarchy and this had negative effects on attitudes. Possibly, because this was a remoter effect rather than an immediate characteristic, it received little spontaneous mention although the direct question showed it was nevertheless a common grievance. A promotional sequence has been described by Blauner as an 'integrator' increasing the degree of personal investment in a job and he takes this point further by arguing that promotion encourages the individual to 'accept the goals of the organization'.[9] While it is important that individual aspirations of this kind are taken into account in the planning of a new plant, it should at the same time be noted that for the steel workers before innovation the job hierarchy was an integrator for the union rather than for the firm.

In general, the autonomy of production crews in the hand mills had many advantages for the individual, although it was easily seen that these were associated with a consensus in the labour force which extended from the primary group into union activities. This area of research might allow a greater insight into these factors on the part of both the unions and management. If as a result of investigation, this area of behaviour is brought more into the open then it might mean that management and unions come into conflict or reach agreement on this explicit dimension as they differ or agree, for example, on wages and salary negotiations.

In Chapter 11 the concept was discussed of the essential variables and it may be worth attempting to define these, to delineate them in detail and then introduce them as part of innovation. The analogy has been made only in the barest outline but gives us a useful way of thinking about behaviour. In engineering systems, the limits of variable movement can be stated and connected explicitly to effective functioning. There are few general statements available about social variables relating to the degree of their permissible movement as there are about the components of physiological and engineering systems, because interpersonal behaviour still has to undergo more precise description and experiment in actual situations.

An approach might be made to the examination of cellular structure by way of the behaviour units which are amenable to measurement. Some of these are shown in Table 27. With measurements of this kind it might be attempted to construct new primary groups, making estimates during planning either by simulation methods or by reference to published findings at an early stage of innovation design.

Even though assessment could be made with only a fair degree of approximation initially, it would be possible consequently to suggest the adoption or alteration of a design proposal from the study of the likely effects for work behaviour and attitudes. This would be a tentative step although one which might in its application eventually yield predictive analysis. The calculation presents many difficulties, but in principle the structure of interacting variables can be derived from the ranges or frequencies such as those listed in Table 27.

How much can be inferred from the quantification of interpersonal contacts? Expressed by frequency, contacts may seem to leave out a crucial body of communication content, but they reveal characteristic patterns of how information is dealt with by the team and of relationships between its members and other personnel. This measure did in fact reveal, as we have seen in Chapter 11, a consistent outline pattern for teams and for the operative positions. Contact frequency and range per hour for each operative position represent the distribution of a social network, the degree of isolation in team positions and, affecting individual performance, any tendency to unequal load of contact (relative underload or overload) on some operatives. In making estimates, two aspects are to be included: the minimal rates of contact necessitated by technical performance, and secondly the rates of further interaction as they might develop from this minimal basis, taking into account the conditions of machine layout and noise. Forecasting can be attempted of what are likely to be the different forms of team structure which would result from these differential conditions. We might also recognize here that certain forms will be more desirable than others to the employees, and therefore the use of attitude surveys can be part of design procedure.

This would mean making an advance analysis, which is *social* i.e. extending to the interconnection between job roles as well as operational to the actual tasks of a given operative. Job analysis which includes the social unit thus

Table 27. The social variables in plant design

The work group *Variables*	*Quantity* *Minimum Maximum*
1. Contact frequency for each operative position (per hour)	O . . . N
2. Contact range (per hour)	O . . . N
3. Number of job exchanges (per hour)	O . . . N
4. Decisions taken by line operatives (per hour/shift)	
(a) Technical	O . . . N
(b) Team organization	O . . . N
5. Relief of other operatives (frequency or duration per hour)	O . . . N
6. Promotion grades	O . . . N
7. Contact (hourly) upwards to supervisors	O . . . N
8. Contact (hourly) upwards to managers	O . . . N
9. Contact ratio, team to supervisors (e.g. relative frequency of contacts *received from,* to contacts *initiated with* supervisors)	$x : y$
10. Contact ratio, team to managers	$x : y$

implies a very different approach from the usual analysis of job elements. Questions like the following would have to be answered: What behaviour appears to be of particular value to the individual? Is contact between operatives likely to be facilitated by machine design? What alternatives would there be in the kinds of network associated with different machine lay-outs?

In Table 27 job exchanges refer first to the task rotation which may be fixed by management; it must also be decided whether an elaboration of exchanges is to be allowed at the discretion of operatives. This has some overlap obviously with the decisions to be allowed to the leading hand or to the team, over job performance and over the task roles of individual members (team organization). These are the quantifiable aspects of the concepts of autonomy and control in the work group.

Rate of contact upwards would reflect the ease of access from the team to supervisors or managers. Calculation of an additional measure of this type of contact, the frequency ratio, gives the degree to which information can be expected to pass easily from one level to the one above, and provides another indicator of autonomy. To revert to a point made previously and which is worth reiterating, one would have to consider how the managerial and supervisory leadership role towards subordinates is to be defined because clearly this will influence how far the team can be flexible in the behaviour followed by its members. Estimates of contact frequency from management levels also make necessary an explicit definition of purposes in behaviour towards subordinates and a consequent reappraisal of the methods of control to be used towards the operating teams.

We have thus begun to consider in a speculative way the range of variables which could be admitted into a plan for innovation. Social and technical factors might conceivably be incorporated together by using a formula based on these elements of group behaviour and stated objectively for application to the design of new technologies. Behind this approach are the assumptions which have been discussed in the previous chapters. We have seen how the work group functioned to bring positive results for individuals. As the converse of this, in electrolytic tinning, both rate and quality of social activities were associated with dissatisfactions. It might be possible now to work towards the definition of structures appropriate for given occupations or industries, bearing in mind the expectations held by employees.

In the steel industry, innovation will continue to occur with intervening short-term periods of stability, and there will be opportunity for the results of investigation at one point of time to be utilized or tested in the next wave of major changes. But innovating policies in the future are unlikely to stop at the boundaries of the firm. Lupton has outlined a code of procedure to include the individual's roles outside as well as inside the workplace:[10] for example, when planning occupational changes, allowance would be made for family circumstances and position in the life-cycle in order to place individuals more suitably into the new job hierarchies. An effective approach for innovation could therefore be one where the use of empirical findings on job skills, individual capacities and retraining methods are combined with attention to behaviour in primary groups and to the broader influences on adjustment.

REFERENCES

1. Louis E. Davis in 'The Design of Jobs', *Industrial Relations,* **6**, No. 1 (1966), p.45, gives a critical discussion of worker communications programmes.
2. *For example,* E. L. Trist, G. Higgin, H. Murray, A. Pollock, *Organisational Choice,* Travistock Publications, London, 1963.
3. *See also* Louis E. Davis and R. Werling, 'Job Design Factors', *Occupational Psychology,* **34**, No. 2 (1960).
4. William A. Faunce, 'Automation and the Automobile Worker', *American Sociological Review,* **23**, No. 4 (1958).
5. Robert Blauner, *Alienation and Freedom: The Factory Worker and His Industry,* University of Chicago Press, 1964.
6. Nigel Walker, *Morale in the Civil Service,* Edinburgh University Press, 1961.
7. Isobel Menzies, 'A Case Study in the Functioning of Social Systems as a defence against Anxiety', *Human Relations,* **13**, No. 2 (1960), p. 97.
8. John H. Goldthorpe, David Lockwood, Frank Bechhofer, Jennifer Platt, *The Affluent Worker,* Cambridge University Press, 1968.
9. Robert Blauner, *Alienation and Freedom: the Factory Worker and His Industry,* University of Chicago Press, 1964, p. 148.
10. Tom Lupton, *A model for Diagnosis and Decision-making in Situations of Technical Change,* Organisation for Economic Co-operation and Development, Paris, November, 1966.

AUTHOR INDEX

SUBJECT INDEX